Blockchain of Things

In recent years, Internet of Things (IoT) has reshaped the whole tech-industry into a smart industry with a capability of intelligent decision-making. However, security, privacy, and interoperability of the connected devices are still the challenges faced by the IoT industry. Blockchain capabilities can help in addressing these challenges. The convergence of Blockchain and IoT named as Blockchain of Things (BCoT) is a major technical breakthrough that provides basic services with additional functionalities, benefits, and implementations in terms of secure, seamless, and autonomous computing environment. The amalgamation of Blockchain and IoT provides benefits in terms of improved and efficient qualities such as interoperability, reliability, scalability, privacy, and security of IoT.

The main objective of this book is to provide academicians, researchers, and practitioners a common platform for understanding current state-of-the-art development in techniques/technologies from Blockchain and its integration with IoT. All the aspects such as theoretical and practical contributions, simulated works, review, and case studies will be available to the reader which will provide them with a comprehensive guideline to extend the state-of-the-art by developing novel methods for providing solutions for the Blockchain of Things. This book:

- Covers comprehensive and deep understanding of usage of Blockchain for powering up IoT systems.
- Discusses the latest advancements in architecture, key technologies, security, privacy, service models, and frameworks for Blockchain of Things.
- Covers Blockchain consensus algorithms specialized for IoT architectures.
- Illustrates key case studies based on inherent integration of Blockchain and IoT in various application domains.

Blockchain of Things

Research Challenges and Applications

Edited by
Sangita S. Chaudhari, Gautam Borkar,
Dhanajay Dakhane, Sachin Jain, and
Abhi Deshmukh

CRC Press
Taylor & Francis Group
Boca Raton London New York

CRC Press is an imprint of the
Taylor & Francis Group, an **informa** business

First edition published 2026
by CRC Press
2385 NW Executive Center Drive, Suite 320, Boca Raton FL 33431

and by CRC Press
4 Park Square, Milton Park, Abingdon, Oxon, OX14 4RN

CRC Press is an imprint of Taylor & Francis Group, LLC

First edition published by CRC Press 2026

ISBN: 9781032954080 (hbk)
ISBN: 9781032954097 (pbk)
ISBN: 9781003584728 (ebk)

DOI: 10.1201/9781003584728

Typeset in Sabon
by Newgen Publishing UK

Contents

Preface

Blockchain technology is an emerging technology that acts with decentralized, distributed, public and real-time ledger to store transactions. A Blockchain is a series of blocks, and each block is linked to its previous blocks. In most of the IoT-enabled applications, IoT nodes generate and transfer sensitive and confidential data. The transactions in Blockchain can be seen as an effective mechanism to transfer data between IoT nodes. The IoT nodes are smart devices with embedded sensors, actuators, programs, and able to communicate with other IoT nodes. The role of Blockchain in IoT is to provide a procedure to process secured records of data through IoT nodes. Being a secure technology, Blockchain can be used publicly and openly and serves to secure communication within IoT nodes in heterogeneous environments.

Although there is tremendous growth in usage of IoT in various applications like smart homes and smart cities, e-health, and distributed intelligence, the technology encounters challenges in establishing complete security and privacy at the sensor level as well as at dissemination level. A decentralized approach is used to connect IoT devices and, therefore, it is very complex to use the standard existing security techniques in the communication among IoT nodes. With the use of Blockchain technology, security in transactions among the IoT devices can be achieved. It provides a decentralized, distributed, and publicly available shared ledger to store the data of the blocks that are processed and verified in an IoT network. The data stored in the public ledger is managed automatically by using the peer-to-peer topology.

This volume comprises 12 chapters. The technical content begins with the fundamentals of Blockchain and IoT and its integration, and the book further includes chapters the discuss various frameworks, algorithms, and case studies. Chapter 1 covers basics of Blockchain and its integration with IoT. Chapter 2 discusses state-of-the-art layered framework for Blockchain of things, benefits, and future trends. Chapter 3 states importance of privacy in Blockchain of things. It also shows the effectiveness of this framework to secure context-aware information in IoT ecosystem. Chapter 4 discusses the basic requirements that need to get fulfilled while integrating Blockchain and also demonstrates the usefulness of such integration for various applications. Chapter 5 covers trust management in Blockchain and IoT and its efficiency on its integration. Chapter 6 is an amalgamation of optimization and BCoT for the development of secure and privacy centric environment. Chapter 7 presents the consensus algorithms and their applicability in IoT environment. Chapter 8 is a Blockchain-based solution to MANET, enhancing overall efficiency of network. Chapter 9 presents a case study highlighting the integration of Blockchain and IoT in healthcare application. Chapter 10 presents an investigation of quick commerce security and applicability of BCoT in these commerce platforms. Chapters 11 and 12 present the benefits, limitations, research challenges, and future trends in Blockchain of Things.

We would like to express our sincere thanks and gratitude to the authors for their outstanding contributions to Blockchain of Things. Without the contributions of the authors, it would have been very difficult to submit this proposal. As editors, we hope that this book will help and stimulate the researchers to pursue research in this domain. Also, we expect the researchers to benefit from the huge set of applications stated in this book. Special thanks goes to our publisher, CRC Press/Taylor and Francis group.

We hope this book will present promising ideas and outstanding research contributions supporting further development of Blockchain of things.

Editor Biographies

Sangita S. Chaudhari obtained her Master of Engineering (Computer Engineering) from Mumbai University, Maharashtra, India, in 2008 and Ph.D. in geographic information system (GIS) and remote sensing from the Indian Institute of Technology Bombay, Mumbai, India, in 2016. Currently, she is working as professor and head of the Department of Computer Science and Engineering, Ramrao Adik Institute of Technology, Nerul, Navi Mumbai. She has published several papers in international and national journals and conferences besides contributing book chapters. Her research interests include image processing, information security, Blockchain technology, GIS, and remote sensing.

Gautam Borkar received his bachelor's degree from the National Institute of Technology, Jalandhar, Punjab, India, and completed his masters and Ph.D. from Sant Gadge Baba Amravati University, Amravati. Presently, he is working as a professor in Ramrao Adik Institute of Technology, Nerul, Navi Mumbai. His current research interests include network security, trust management, Blockchain and security in wireless sensor networks. He had published several papers in international journals.

Dhanajay Dakhane is currently working as a professor in the Department of Computer Science and Engineering, Ramrao Adik Institute of Technology, Nerul, Navi Mumbai. He has completed his bachelor, masters and Ph.D. in computer science and engineering from Sant Gadge Baba Amravati University, Amravati. His research interest includes covert channels and its counter measures, secret and multi-party computations. His research works are published in international and national journals. He has around 90+ citations to his research work, out of which one is i-06 indexed and four are h-indexed.

Sachin Jain is working as an assistant professor in the Computer Science Department, Oklahoma State University, Stillwater, Oklahoma. He has completed his bachelor's in computer technology and master's in computer science and engineering from Rashtrasant Tukadoji Maharaj Nagpur University, Nagpur, India and Ph.D. in computer science and engineering from Sant Gadge Baba Amravati University, Amravati, India. Dr. Sachin Jain has more than 18 years of teaching experience and his key research areas are wireless sensor networks, artificial intelligence, machine learning, Internet of Things, and image processing. He is a member of International Association of Engineers (IAENG), Computer Society of India (CSI), and Indian Society for Technical Education (ISTE). He has actively participated in designing various courses at the undergraduate and postgraduate level. He has guided more than 50 undergraduate projects and more than ten postgraduate research projects. He has published more than 20 research papers in international conferences and journals. He has around 30 citations to his research work, out of which one is i-10 indexed and four are h-indexed. He is a reviewer for many international conferences and *IEEE Transactions on Cybernetics*.

Abhi Deshmukh has extensive experience in product engineering, sales, and business management across multiple domains like analytics, AI/machine learning, enterprise marketing management, Cloud and SaaS solutions, IT service management, workflow automation, business support system/operating support system (BSS/OSS), usage chargeback, revenue settlement and billing, computer telephony integration, embedded, real-time applications, and network protocol stack development. Abhi is the CTO of iHub Anubhuti-IIITD Foundation, New Delhi, India. He led a Technology Innovation Hub (TiH) at IIIT-Delhi established by the Department of Science and Technology (DST), Government of India, under the National Mission on Interdisciplinary Cyber-Physical Systems (NM enhanced survey information synthesis Interdisciplinary Cyber Physical System (ICPS), www.nmicps.in). His primary focus lies in developing technologies in the areas of cognitive computing, computer vision, conversational AI, computational lingusitics (natural language processing (NLP)), climate, MedTech, healthcare, and social sensing. He also has led efforts for startup incubation, investments, and grants for research projects. He has been granted three US patents and has one patent is pending with the United States Patent Office. Abhi holds a bachelor's degree in computer science and engineering from the Amravati University, India, a masters (MS) degree in software engineering from Rensselaer Polytechnic Institute (RPI), Troy, New York.

Contributors

Sonal Bankar, Lokmanya Tilak College of Engineering, Koparkhairane, Navi Mumbai, Maharashtra.

Smita Bharne, Ramrao Adik Institute of Technology, D. Y. Patil Deemed to be University, Nerul, Navi Mumbai, Maharashtra.

Gautam M. Borkar, Ramrao Adik Institute of Technology, D. Y. Patil Deemed to be University, Nerul, Navi Mumbai, Maharashtra.

Smita Dange, Fr. C. Rodrigues Institute of Technology, Vashi, Navi Mumbai, Maharashtra.

Jagjit Singh Dhatterwal, School of Computer Science and Artificial Intelligence, SR University, Warangal, Telangana.

Nitesh Ghodichor, Priyadarshini College of Engineering, Computer Technology, Nagpur, Maharashtra.

Vaishali Ghodichor, G. H. Raisoni College of Engineering and Management, Nagpur, Maharashtra.

K.V. Haneefa, School of Commerce, Finance and Accountancy, Christ (Deemed to be University), Bengaluru, Karnataka.

Poonam Samir Jadhav, D. J. Sanghavi College of Engineering, Mumbai, Maharashtra.

Bharti Joshi, Ramrao Adik Institute of Technology, D.Y. Patil Deemed to be University, Nerul, Navi Mumbai, Maharashtra.

Kuldeep Singh Kaswan, School of Computer Science and Engineering, Galgotias University, Greater Noida.

Shaila P. Kharde, Shreeyash College of Engineering and Technology, Chh. Sambhaji Nagar, Maharashtra.

Madhumitha Kulandaivel, SRM University, Kattankulathur, Chennai, Tamil Nadu.

Suhas Lawand, Xavier Institute of Engineering, Mumbai, Maharashtra.

Jayant R. Mahajan, Shreeyash College of Engineering and Technology, Chh. Sambhaji Nagar, Maharashtra.

Kiran Malik, GL Bajaj Institute of Technology and Management, Greater Noida.

Keerthana N.V., Velalar College of Engineering and Technology, Erode, Tamil Nadu.

Prashant Nitnaware, Pillai College of Engineering, Panvel, Maharashtra.

Prema P., Velalar College of Engineering and Technology, Erode, Tamil Nadu.

Puja Padiya, Ramrao Adik Institute of Technology, D. Y. Patil Deemed to be University, Nerul, Navi Mumbai, Maharashtra.

Vedant Ganesh Padwal, Lokmanya Tilak College of Engineering, Koparkhairane, Navi Mumbai, Maharashtra.

Arvind Panwar, School of Computer Science and Engineering, Galgotias University, Greater Noida.

Deepa Parasar, Amity School of Engineering and Technology, Amity University, Mumbai, Maharashtra.

M. Parimala Devi, Velalar College of Engineering and Technology, Erode, Tamil Nadu.

R.S. Shudapreyaa, Kongu Engineering College, Erode, Tamil Nadu.

Kaikashan I. Siddavatam, Lokmanya Tilak College of Engineering, Koparkhairane, Navi Mumbai, Maharashtra.

Harmandeep Singh, School of Commerce, Finance and Accountancy, Christ (Deemed to be University), Bengaluru, Karnataka.

Shraddha C. Subhedar, Saraswati College of Engineering, Mumbai University, Kharghar, Maharashtra.

Amarsinh V. Vidhate, Ramrao Adik Institute of Technology, D. Y. Patil Deemed to be University, Nerul, Navi Mumbai, Maharashtra.

Rajesh D. Wagh, Shreeyash College of Engineering and Technology, Chh. Sambhaji Nagar, Maharashtra.

Ojas Arun Yadav, Lokmanya Tilak College of Engineering, Koparkhairane, Navi Mumbai, Maharashtra.

Sulbha Yadav, Lokmanya Tilak College of Engineering, Koparkhairane, Navi Mumbai, Maharashtra.

Chapter 1

Everything about Blockchain technology and IoT integration with a keen eye

Rajesh D. Wagh, Jayant R. Mahajan, and Shaila P. Kharde

1.1 INTRODUCTION

Secured chains of blocks are not a novel notion. In 1991, Stuart Haber et al. introduced it as a way to digitally timestamp electronic documents in order to guard against tampering [1–3]. Bitcoin, one of the most popular cryptocurrencies, has seen tremendous growth, with its capital market valued at $10 billion in 2016 [4]. Transactions in the Bitcoin network could occur without the involvement of a third party with a specifically created data storage structure. The Blockchain, which was first developed in 2008 and put into use in 2009, is the fundamental technology behind Bitcoin [5,6]. Bitcoin is a digital currency built on Blockchain technology that allows for unrestricted global trading without the need for a banking institution to guarantee it. However, Bitcoin is only a monetary implementation of Blockchain technology. Blockchain is described as an immutable auditable timestamp and tamper-resistant block ledger that shares, stores, and uses information in a peer-to-peer (P2P) manner [7–11]. Additionally, there are various types of Blockchain according to their unique features and applications: consortium Blockchain, also known as "federated Blockchain," fall between private and public Blockchain in terms of management and permissions. Public Blockchains are fully decentralized, allowing anyone to join the network and participate in its management; private Blockchains are exclusive to invited members of a single organization; and third, there are alliance Blockchains [3].

The Internet of Things (IoT) is a network of physical objects that are linked together over the internet in order to gather data and make deft decisions. Globally, this field is expanding quickly, and research related to it is also expanding quickly. By 2025, there will be 25 billion IoT devices installed, predicts Gartner, the global research and advisory firm [12,13]. By 2025, the estimated global economic effect of IoT devices would be between $2.7 and $6.2 trillion [13,14]. But as the number of IoT devices rises dramatically, so do the security risks associated with them. Numerous issues, including data provenance binding, secure data storage, fault tolerance, data origin

DOI: 10.1201/9781003584728-1

integrity, confidentiality, chain integrity, replay assaults, and data availability, confront the current IoT systems [13]. Even though the problems confronting the IoT paradigm are diverse, there is some correlation between them. These challenges range from intricate privacy and security problems to fundamental network architecture problems. As a result, the obvious necessity to comprehend these issues and find solutions has spurred several studies to investigate and offer suitable answers. Blockchain is a cutting-edge and new technology that has been shown to be an essential remedy for IoT problems.

The remainder of the chapter can be summed up as follows: The chapter's introduction and an explanation of Blockchain technology are presented in Section 1.1. The architecture of IoT and its challenges are presented in Section 1.2. The characteristics of Blockchain technology are introduced in Section 1.3. The classification of Blockchain technology based on access and control mechanisms is presented in Section 1.4. The consensus algorithms used in Blockchain technology in IoT are presented in Section 1.5. The use cases of Blockchain technology are presented in Section 1.6. Finally, the conclusion is presented in Section 1.7.

1.2 IOT ARCHITECTURE

As seen in Figure 1.1, a basic IoT architecture is composed of three layers. The application layer, Server/Cloud, gateways, and devices are the main parts of the Internet of Things architecture. The sensor devices, often referred to as the physical layer, are part of the perception layer. The data is gathered by the devices/sensors and sent to the network layer. The network layers that link the application layer and perception layer. Application layer receives data, analyzes it, processes it, and then sends the required service back to the physical layer devices via network layer [15].

1.2.1 Challenges of the IoT Centralized Model

The following are some of the issues related to centralized IoT architecture:

1. *Scalability*: Because Blockchain technology is still in its infancy and has scalability problems, handling massive volumes of IoT data can be challenging [13,16,17].
2. *Interoperability*: It can be difficult to integrate different Blockchain platforms and IoT systems because of their disparate communication protocols [13,18].
3. *Security and privacy*: IoT devices are susceptible to cyber-attacks, and the chance of smart contract vulnerabilities rises when they are integrated with Blockchain technology. Because of the centralized IoT paradigm, data processing and storage are handled by a single

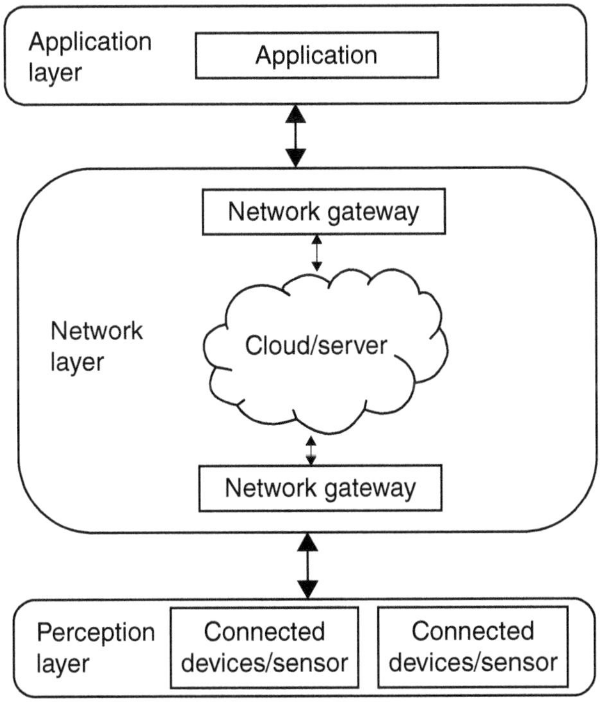

Figure 1.1 Three-layered IoT architecture [15].

database, raising the possibility of security risks like denial of service (DoS) attacks as well as challenges to the integrity and privacy of the data [13,15,16,18].

4. *Data management*: Managing and storing the massive volumes of data generated by IoT on a Blockchain can be challenging.
5. *Regulatory frameworks*: Widespread adoption of Blockchain and IoT integration is hampered by unclear regulatory frameworks.
6. *Energy consumption*: Blockchain technology uses a lot of energy, which may be problematic for IoT gadgets that run on batteries [13,18].
7. *Standardization*: To ensure smooth integration, protocols and communication interfaces must be standardized.
8. *Latency*: Real-time IoT applications may be affected by the slowness of Blockchain technology.
9. *Cost*: Businesses may find it difficult to embrace due to the significant integration and implementation costs. The price of maintaining a centralized server and database with highly capable hardware and software [15].

10. *Single point of failure*: The entire IoT system fails if a server fails [15].
11. *Scalability*: When there are fewer IoT devices, the centralized IoT approach performs well. As the number of IoT devices develops, scalability problems occur as the network becomes more complicated [15].
12. *Diversity and accessibility*: Accessing the network can be challenging due to heterogeneous IoT devices coupled to a centralized IoT model [15].

1.3 BLOCKCHAIN (BC) ARCHITECTURE

The technology known as Blockchain makes it possible for a collection of dubious individuals to verify transactions. It offers a distributed, secure, transparent, auditable, and immutable ledger [16]. Although the Blockchain was first presented in 2008, it wasn't put into practice until 2009. It can keep track of and save data from a big number of nodes, the distributed ledger technology, also known as shared ledger technology, and is based on the Blockchain. A block has just one parent block if the block header contains the previous block hash. It's important to remember that hashes of uncle blocks—children of the block's ancestors—would likewise be kept on the Ethereum network. The genesis block, which is the initial block in a Blockchain without a parent block [13]. As seen in Figure 1.2, a block is made up of the block header and the block content. Specifically, the block header consists of [19]:

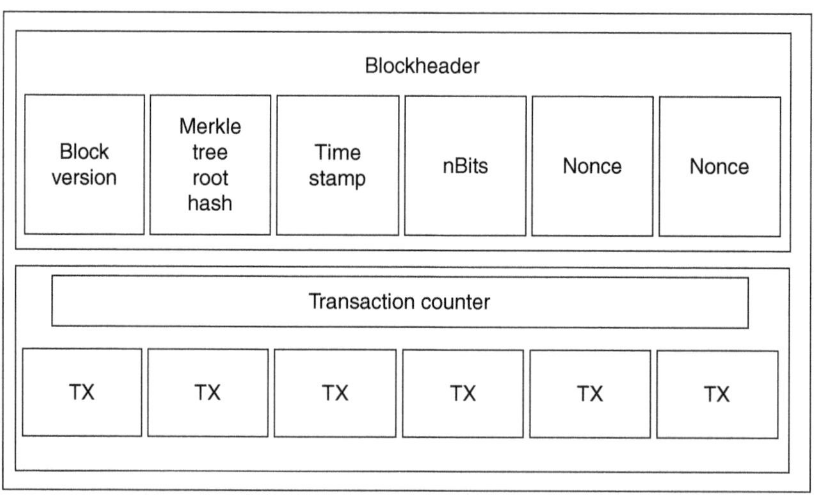

Figure 1.2 Block structure.

(i) Block version: designates the block validation rules to be adhered to.
(ii) The hash value of every transaction in the block is the Merkle tree root hash.
(iii) Timestamp: the moment as of January 1, 1970, expressed in universal time in seconds.
(iv) nBits: The target threshold of a valid block hash.
(v) Nonce: A 4-byte field that typically begins at 0 and grows for each block.

1.3.1 Characteristics of Blockchain

The primary differentiators that enhance any industry utilizing Blockchain technology are as follows [20–23]:

- *Decentralization*: Without the oversight of a central body, network users have access to data records.
- *Distribution*: A copy of the data records, which are updated continuously, is held by each node.
- *Security*: The created blocks can't be removed or altered thanks to the Blockchain structure of linked blocks using a hash technique.
- Automation is achieved through the use of smart contracts, which allow a computer to automatically start a certain operation when a set of predetermined criteria are met.
- *Traceability*: Since its founding, Blockchain has maintained a historical record of every data. One can link a record like this to the initial action.
- *Privacy*: Using a private/public key, participants' information is kept anonymous, even if Blockchain technology is transparent.
- *Reliability*: Because of its advantages and sturdy structure, Blockchains have been effectively adopted by a number of companies.

1.4 CLASSIFICATION OF BLOCKCHAIN

1.4.1 According to the Access Mechanisms

Table 1.1 illustrates the three classifications of Blockchains based on access mechanisms: consortium, private, and public [24,25]. The transparent transaction history stored on a public Blockchain is accessible to everybody who joins the network. Both the identities of network participants and the transactions on the public Blockchain are anonymous. In a private Blockchain, access permissions and network regulations are set by a single entity, giving it control. Although there may be some decentralization among the numerous nodes running the Blockchain that the owner controls, trust is centralized at the owner. Although multiple nodes administering the

Table 1.1 Blockchain classification according to access mechanisms [26]

	Public	*Private*	*Consortium*
Network structure	Decentralized	Centralized	Partially decentralized
Controlled by	All network participants	Trusted entity (Blockchain owner)	Predetermined group of network participants
Efficiency	Low	High	Medium
Security	Higher due to distribution	Lower due to centralization	Average due to partial distribution
Privacy	Low—all transactions are transparent	High—access to data is controlled by the trusted entity	Medium—access to data is controlled by a group of network participants
Use case examples	Crypto currency, Bitcoin, Ethereum, Litecoin, etc.	Company-owned Blockchains, government applications	Consortium of companies, multiple government agencies, Hyperledger, Quorum, Corda, Ripple, etc.

Table 1.2 Control- and access-based Blockchain classification [26]

		Permissionless			*Permissioned*	
	Read	*Write*	*Join*	*Read*	*Write*	*Join*
Public	Any	Any	Any	Any	Authorized nodes	Any
Private	Authorized nodes	Authorized nodes	Authorized nodes	Authorized nodes	Operator	Authorized nodes
Consortium	Authorized nodes	Authorized nodes	Authorized nodes	Authorized nodes	Consortium validators	Authorized nodes

Blockchain that are under the owner's control may exhibit some degree of decentralization, trust is concentrated at the owner. Consortium Blockchains are designed for applications in which a number of players engage with one another; a predefined set of network users oversees the block chain's maintenance and consensus process [26].

1.4.2 According to Control Mechanisms

Blockchains can be categorized as permissionless or permissioned based on the controls they use. Table 1.2 associates access-based classifications with permissioned and permissionless Blockchains. Any node can join a permissionless Blockchain and take part in originating and confirming transactions, strengthening the consensus process. Blockchains with permissioned participation: The nodes that take part in the Blockchain are pre-selected and granted authorization to do so. An organization or collection of organizations can record communications, events, and transactions in an unchangeable way using permissioned Blockchains. A single organization or a collection of organizations controls the Blockchain, and the degree of decentralization varies depending on how the network interacts [26,27].

1.5 CONSENSUS ALGORITHMS FOR BLOCKCHAINS IN IOT

The architecture in which the consensus will be utilized, the hardware specifications, and the attack vector that is meant to be mitigated are the three key determinants of its acceptance. As a result, when selecting the consensus technique, the quantity of nodes and the processing overhead are crucial factors to take into account. An overview of the consensus algorithms pertaining to Blockchain access, Blockchain control methodology, and advantages and disadvantages of IoT is provided in Table 1.3 [25–27].

Table 1.3 Overview of consensus algorithms for Blockchains in IoT [26]

Consensus algorithm	*Access*	*Control*	*Positive aspects for IoT*	*Negative aspects for IoT*
PoW	Public	Permissionless	Consensus was reached with few messages exchanged.	high usage of energy and computing
PoS	Public	Permissionless	Scalable, lower power consumption	Overloadinfewnodes can impact in the operation of the Blockchain
PoSpace	Public	Permissionless	Lower power consumption	Requires high amount of memory/storage
PBFT	Private/ Consortium	Permissionless	Less hardware/energy requirements	Not scalable
dBFT	Private/ Consortium	Permissioned	Scalable and less hardware/energy requirements	Problems in dynamic scenarios
FBA	Private/ Consortium	Permissioned	Scalable and less hardware/energy requirements	It is required that "important"nodesare trusted
IBFT	Private/ Consortium	Permissioned	Less hardware/energy requirements	Not scalable, produces empty blocks
RAFT	Private/ Consortium	Permissioned	Less hardware/energy requirements	Not scalable, serialization of requests

Proof-of-Work (PoW) regulates the creation of new blocks by requiring the solution of a resource-intensive cryptographic puzzle. Miners must typically locate a nonce value to add in the block so that the new block's hash is less than a target value in order to complete a PoW task. To further regulate the block generation pace, the difficulty can be changed over time. The PoW algorithm has an alternative called Proof-of-Stake (PoS). PoS uses a random selection of nodes based on coin aging or wealth in order to lessen the difficulty of the block creation operation [25,26,28].

In contrast to PoW, Proof-of-Space (PoSpace) was suggested as a more energy-efficient alternative. PoSpace can concentrate on either temporary or permanent space. PoSpace typically employs proof of safe erasure functions or memory-hard functions (MHF), which necessitate memory- and space-intensive computations. This method has the benefit that a verifier can verify the findings provided by the node that produced the block with little processing and space requirements [25,26,29].

When a portion of the nodes in a distributed network are malevolent or malfunctioning, the network can nevertheless reach consensus correctly, a phenomenon known as practical byzantine fault tolerance (pBFT). pBFT requires 3f +1 nodes to correctly attain consensus when there are f defective nodes in the network. A leader node is chosen upon the creation of a new block. The consensus process is then initiated by the leader node by forwarding the block to the network's active validation nodes for approval. The block gets attached to the Blockchain if more than two-thirds of the active validation nodes vote in favor of validating it [26,30].

Similar to pBFT, delegated byzantine fault tolerance (dBFT) uses votes to reach consensus on new information. On the other hand, in the dBFT, the requester elects validators—nodes that both validate and vote—for every consensus. For the subsequent consensus process, the requester may designate a different node as validator if they have doubts about the selected validator [26].

Federated byzantine agreement (FBA), which is based on a set of affirmative votes, forms consensus similarly to pBFT. Every node is aware of a group of other nodes that it has predetermined to be "important" based on subjective standards like credit standing. When carrying out the consensus, the new information is validated by the important nodes once the requester node starts the consensus procedure [26].

In order for Istanbul byzantine fault tolerance (IBFT) to work, the new data must be validated by more than two-thirds of the Blockchain's active nodes. However, a "round robin" method may be used to choose the proposer—the node that initiates or directs the consensus process. The consensus is initiated by the proposer node without a leader needing to be selected. IBFT is regarded as a pBFT adaptation that enables various nodes to continuously generate new blocks [26, 31].

Along with having a leader who initiates the validation, RAFT also requires the network to be configured with 2f +1 nodes in order to be able to withstand f malfunctioning nodes. The time it takes to switch the leader is arbitrary, though, and it does not generate empty blocks like IBFT does. For the purpose of performing validation, the leader responds to all node requests and forwards them to all followers, or other nodes in the network [26,32].

1.6 USE CASES FOR BLOCKCHAIN AND IOT INTEGRATION

Here are several industries where Blockchain and IoT integration has demonstrated significant potential, here are some examples:

1.6.1 Supply Chain and Logistics

In supply chain management, IoT devices (such as radiofrequency identification (RFID) tags and global positioning system (GPS) trackers) can monitor the movement and condition of goods. Blockchain can record every transaction (e.g., shipping, customs checks, and handling) in a secure, transparent, and immutable ledger. A strategy for incorporating Blockchain and IoT technology into contemporary supply chains is presented by Nejc Rožman et al. [33]. In order to address the complexity of contemporary supply chains, they suggest a novel logistics platform that is constructed as a dispersed network of nodes and divides them into smaller, independently operating components. The Blockchain and IoT-based pharmaceutical supply chain project, as presented by Mohammad Zayd et al. [34], is an endeavor that seeks to enhance the current pharmaceutical supply chain models through the use of cutting-edge technologies like Blockchain and IoT.

1.6.2 Healthcare

In healthcare, IoT devices such as wearables and medical sensors collect sensitive patient data. Blockchain can provide a secure and transparent way to store this data while ensuring that only authorized parties can access it. According to Helmy et al. [35], Alzimio is a smartphone software that uses geofencing and activity-based alerts to assist people with dementia, autism, and Alzheimer's disease. Using a variety of threshold-based algorithms, including "max-in-window," which produces a delay of less than 30 seconds, the app has demonstrated an ability to detect activities with an accuracy rate of over 95 percent and little latency. According to authors A. Celesti et al. [36], Cloud computing, Blockchain, and IoT layers are used in this work, which introduced a telemedical laboratory service. Instantaneous communication between clinical tests performed on patients utilizing IoT medical devices demonstrates a distributed approach to healthcare.

1.6.3 Automotive and Autonomous Vehicles

Autonomous vehicles (AVs) rely heavily on IoT sensors for navigation and safety. Blockchain technology can be used to track the data generated by these sensors and ensure that it is not tampered with. Moreover, Blockchain can enable secure vehicle-to-vehicle (V2V) and vehicle-to-infrastructure (V2I) communication. By integrating Blockchain technology into intelligent transport systems (ITS) or individual vehicle units, Nidhee Kamble et al. [37] propose ways to advance the autonomous vehicles (AV) industry. The study by Stefan Iordache et al. [38] explores how Blockchain technology might strengthen AV security by protecting vehicle communication networks. This strategy seeks to increase pedestrian and driver safety, secure vehicle peer-to-peer communication, and lessen traffic congestion by combining Blockchain technology with AI-based prediction algorithms.

1.7 CONCLUSION

Blockchain's four key features—audibility, anonymity, persistency, and decentralization—have shown that they may transform well-established sectors. The potential for Blockchain technologies to revolutionize IoT applications in terms of enhanced security, decentralized communications and control, transparency, traceability, and reliability, as well as new forms of trust, is still in its early stages of development, pilot programs, and demos. We provide a thorough introduction to Blockchain and IoT in this chapter. We first provide an introduction of the IoT architecture and talk about some of the problems with the centralized IoT architecture. After that, a summary of Blockchain technology is covered, including its architecture, salient features, and classification. Next, we talk about common Blockchain consensus algorithms. Blockchain-based apps are becoming more and more popular these days, and we intend to look into them in-depth in the future.

REFERENCES

[1] J. R. Anderson, *Security Engineering: A Guide to Building Dependable Distributed Systems*, 2nd ed. Indianapolis, IN, USA; Wiley, 2008.

[2] L. Lamport, "Password authentication with insecure communication," *Communications of the ACM*, vol. 24, no. 11, pp. 770–772, Nov. 1981.

[3] M. N. Mumtaz Bhutta et al., "A survey on blockchain technology: Evolution, architecture and security," *Special Section on Intelligent Big Data*

Analytics for Internet of Things, Services and People, IEEE Access, vol. 9, pp. 61048–61073, 2021.
[4] G. Hileman, "State of blockchain q1 2016: Blockchain funding overtakes Bitcoin," 2016. [Online]. Available: www.coindesk.com/state-of-Blockchain-q1-2016/
[5] S. Nakamoto, "Bitcoin: A peer-to-peer electronic cash system," 2008. [Online]. Available: https://bitcoin.org/bitcoin.pdf
[6] A. Baliga, et al., "Performance evaluation of the Quorum blockchain platform," *arXiv:1702.03068, arXiv:1809.03421,* 2018.
[7] Z. Zheng et al. "An overview of blockchain technology: Architecture, consensus, and future trends," In: *IEEE 6th International Congress on Big Data*, 2017.
[8] Y. Wang et al., "Making sense of Blockchain technology: How will it transform supply chains?" *International Journal of Production Economics*, vol. 211, pp. 221–236, May 2019.
[9] A. Al Sadawi et al., "A survey on the integration of blockchain with IoT to enhance performance and eliminate challenges," *IEEE Access*, vol. 9, pp. 54478–54497, 2021.
[10] J. F. Galvez et al., "Future challenges on the use of blockchain for food traceability analysis," *Trends in Analytical Chemistry*, vol. 107, pp. 222–232, Oct. 2018.
[11] S. Dziembowski, et al., "Proofs of space," In: *Advances in Cryptology—CRYPTO 2015*, pp. 585–605, 2015.
[12] M. Castro, and B. Liskov, "Practical Byzantine fault tolerance," In: *Proceedings of the Third Symposium on Operating Systems Design and Implementation (OSDI'99)*, pp. 173–186. USENIX Association, Berkeley, CA, USA, 1999.
[13] D. Ongaro, and J. Ousterhout, "In search of an understandable consensus algorithm," In: *Proceedings of the 2014 USENIX Conference on USENIX Annual Technical Conference, USENIX ATC'14*, pp. 305–320. Berkeley, CA, USA, 2014.
[14] L. D. Xu et al., "Embedding Blockchain technology into IoT for security: A survey," *IEEE Internet of Things Journal*, vol. 8, no. 13, pp. 10452–10473, 01 July 2021.
[15] A. Reyna at al., "On Blockchain and its integration with IoT. Challenges and opportunities," *Future Generation Computer Systems*, vol. 88, pp. 173–190, Nov. 2018.
[16] H-N. Dai et al., "Blockchain for Internet of Things: A survey," *IEEE Internet of Things Journal*, vol. 6, no. 5, pp. 8076–8094, Oct. 2019.
[17] U. Kumar, M. S. Kaswan, R. Kumar, R. Chaudhary, J. A. Garza-Reyes, R. Rathi, and R. Joshi, "A systematic review of Industry 5.0 from main aspects to the execution status," *The TQM Journal*, pp. 1526–1549, 2023.
[18] H. F. Atlam, M. A. Azad, A. G. Alzahrani, and G. Wills, "A review of Blockchain in Internet of Things and AI," *Big Data and Cognitive Computing,* vol. 4, p. 28, 2020.
[19] M. D. J. Peters, C. M. Godfrey, H. Khalil, P. McInerney, D. Parker, and C. B. Soares, "Guidance for conducting systematic scoping reviews,"

International Journal of Evidence-Based Healthcare, vol. 13, pp. 141–146, 2015.

[20] M. Abouali, K. Sharma, O. Ajayi, and T. Saadawi, "Blockchain framework for secured on-demand patient health records sharing," In: *Proceedings of the 2021 IEEE 12th Annual Ubiquitous Computing, Electronics & Mobile Communication Conference (UEMCON)*, New York, NY, USA, pp. 35–40, 1–4 Dec. 2021.

[21] M. A. Obaidat, M. Rawashdeh, M. Alja'afreh, M. Abouali, K. Thakur, and A. Karime, "Exploring IoT and blockchain: A comprehensive survey on security, integration strategies, applications and future research directions," *Big Data and Cognitive Computing*, vol. 8, p. 174, 2024.

[22] E. A. Shammar et al., "A survey of IoT and blockchain integration: Security perspective," *IEEE Access*, vol. 9, pp. 156115–156150, 2021.

[23] V. Dedeoglu et al., "Blockchain technologies for IoT," *ResearchGate*, Sep. 2019.

[24] T. Alam, "Blockchain and its role in the Internet of Things (IoT)," *International Journal of Scientific Research in Computer Science, Engineering and Information Technology*, vol. 5, no. 1, pp. 151–157, 2019.

[25] X. Wu et al., "M2M blockchain: The case of demand side management of smart grid," In: *2017 IEEE 23rd International Conference on Parallel and Distributed Systems (ICPADS)*, pp. 810–813, Dec. 2017.

[26] M. A. Ferrag, M. Derdour, M. Mukherjee, A. Derhab, L. Maglaras, and H. Janicke, "Blockchain technologies for the Internet of Things: Research issues and challenges," *IEEE Internet of Things Journal*, vol. 6, pp. 2188–2204, 2019.

[27] F. Casino T. K. Dasaklis, and C. Patsakis, "A systematic literature review of blockchain-based applications: Current status, classification and open issues," *Telematics and Informatics*, vol. 36, pp. 55–81, Mar. 2019.

[28] Gartner, "Gartner identifies top 10 strategic IoT technologies and trends 2018. www.gartner.com/en/ newsroom/pressreleases/2018-11-07-gartner-identifies-top-10-strategic-iot-technologies-and-trends. Accessed 10 Jan 2020.

[29] A. Al-Fuqaha, M. Guizani, M. Mohammadi, M. Aledhari, and M. Ayyash, "Internet of Things: A survey on enabling technologies, protocols, and applications," *IEEE Communication Surveys & Tutorials,* vol. 17, no. 4, pp. 2347–2376, 2015.

[30] S. Zafar et al., "Integration of blockchain and Internet of Things: Challenges and solutions," *Annals of Telecommunications*, pp. 13–32, 2021.

[31] A. S. Mohammad et al., "Integration of IoT and blockchain," *Technium*, vol. 3, no. 8, pp. 32–41, 2021.

[32] V. Gazis et al., "Short paper: IoT: Challenges, projects, architectures," *18th International Conference on Intelligence in Next Generation Networks, IEEE*, 2015.

[33] N. Rožman, R. Vrabič, M. Corn, T. Požrl, and J. Diaci, "Distributed logistics platform based on Blockchain and IoT," *Procedia CIRP*, vol. 81, pp. 826–831, 2019.

[34] M. Zayd, A. Johari, and M. Manimozhi, "Supply chain management using Blockchain and IoT in pharmaceutical industries," In: *2023 Innovations in Power and Advanced Computing Technologies (i-PACT)*, IEEE, 2013.

[35] J. Helmy, and A. Helmy, "The Alzimio App for dementia, autism and Alzheimer's: using novel activity recognition algorithms and geofencing," In: *2016 IEEE International Conference on Smart Computing*, June 2016.

[36] A. Celesti, A. Ruggeri, M. Fazio, A. Galletta, M. Villari, and A. Romano, "Blockchain-based healthcare workflow for tele-medical laboratory in federated hospital IoT clouds," *Sensors (Basel)*, 2020.

[37] N. Kamble, R. Gala, R. Vijayaraghavan, and E. Shukla, "Using blockchain in autonomous vehicles," In: *Artificial Intelligence and Blockchain for Future Cybersecurity Applications*, editors: Yassine Maleh, Y.Baddi, M. Alazab, L.Tawalbeh, and I. Romdhani, Springer Cham, pp. 285–305, May 2021.

[38] C. C. Patilea, S. Iordache, and C. Paduraru, "Enhancing autonomous vehicle safety with blockchain technology: Securing vehicle communication and AI systems," *Future Internet*, MDPI, vol. 16(12), pp. 1 20, 2024.

Chapter 2

Review of Blockchain of Things

Benefits, research challenges and future trends

Sulbha Yadav and Sonal Bankar

2.1 INTRODUCTION TO BLOCKCHAIN OF THINGS

The Internet of Things (IoT) developed quickly in terms of resources and grew at an explosive rate, unlike any other technology. The IoT system consists of hundreds of connected and functional gadgets and devices. However, as more devices were connected, additional issues arose, including how to protect data security and manage privacy. It might be necessary to use a Blockchain that is impenetrable in order for it to function well.

By combining the two technologies, the Blockchain of Things (BCoT) seeks to mitigate the vulnerabilities of IoT while offering a secure, independent machine-to-machine communication system [1]. This builds trust and gives BCoT a solid basis that enables it to be used for all kinds of industrial applications. There is a lot of potential to improve security, privacy, and data management in IoT systems by combining Blockchain technology with IoT, or BCoT [2–4]. BCoT is a promising field for further development and research because of the advantages of a decentralized, secure, and effective data management system, even though threats like resource limitations and interoperability must be addressed. This chapter focuses on the advantages and disadvantages of the BcoT as it develops over time.

2.2 BLOCKCHAIN OF THINGS STATE-OF-THE-ART

2.2.1 Internet of Things (IoT)

IoT devices use sensors like geographic positioning system (GPS), light-dependent resistor (LDR), and temperature sensors to gather data from their many users and environmental consumptions. Servers send this data to the cloud for analysis via Bluetooth, local area network (LAN), or WiFi. Cloud software is used to generate insights from data, such as weather analysis (Figure 2.1).

DOI: 10.1201/9781003584728-2

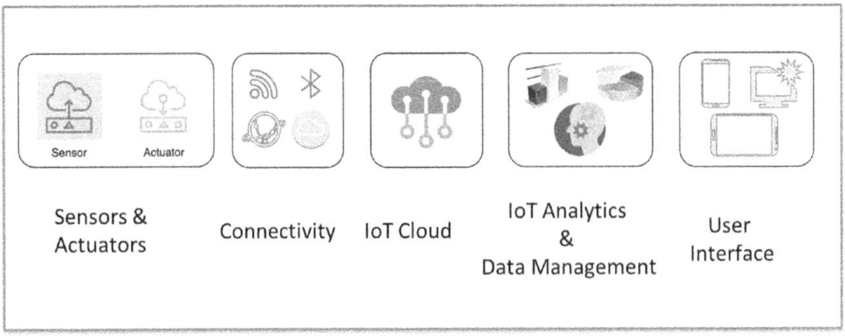

Figure 2.1 Components of IoT.

2.2.2 Blockchain technology

Blockchain is a decentralized computer network that functions as an encrypted database . Since editing or hacking the information would require breaking into millions of computers worldwide, the nature of construction makes it nearly impossible. Blocks of data are connected by timestamps, transaction information, and hashing. By verifying the information before it is added to the chain, miners also make sure that it is permanently stamped, making abuse impossible. A Blockchain platform usually consists of five parts: a virtual machine, a consensus algorithm, a distributed ledger, a peer-to-peer (P2P) network, and encryption (Figure 2.2).

2.2.3 Emergence of BlockChain of Things

The proposed IoT-Blockchain architecture's in literature shows how the various system tiers that integrate Blockchain and IoT technologies operate (Figure 2.3). Increased security, transparency, and automated functions in a number of processes could be possible with this architecture [5–13].

1. *Perception layer* (*data collection*): The foundation of the IoT–Blockchain architecture is the perception layer, which is in charge of gathering information from the physical world using sensors and actuators.

Sensors: Physical phenomena like temperature, humidity, motion, light, and pressure are all recorded by sensors. For instance, the motion and temperature sensors in a smart home will gather information from the surroundings.

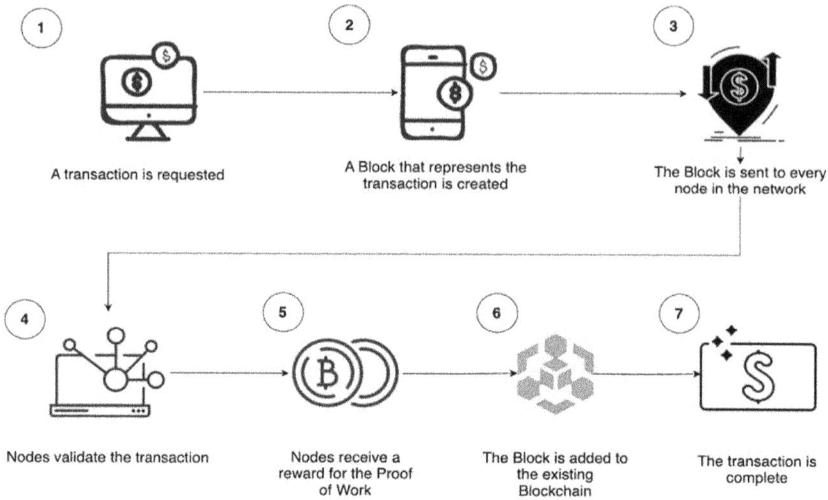

Figure 2.2 Working of Blockchain technology.

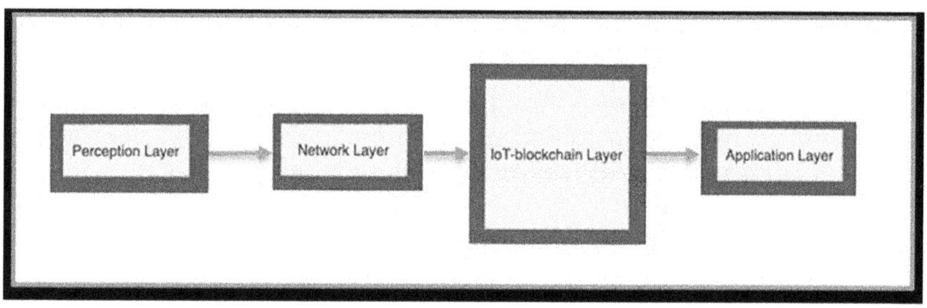

Figure 2.3 Components of Blockchain of Things.

Actuators: Actuators are gadgets that react instantly to control signals; examples include door locking, thermostat adjustment, and lighting. They enable the IoT system to use gathered data to make decisions in real time. Hundreds of thousands of Internet of Things (IoT) devices may be present in this layer, gathering raw data that is sent to the following layer for processing.

2. *Network layer* (secure data transmission): After being gathered by the perception layer's sensors and actuators, the data is transmitted to the network layer. It is a crucial layer that ensures the security, dependability, and efficiency of communication in gateways, IoT devices, and other layers of this architecture.

- Data transmission allows the network layer to send and receive data using a variety of methods, including satellite networks, Wi-Fi, ZigBee, Bluetooth, 5G, and more—anything that would be compatible with an IoT system.
- *Security issues*: The security of the data during transmission is a major concern for this layer. Through encryption, authentication, and the use of only secure protocols like Transport Layer Security/Secure Sockets Layer (TLS/SSL) or Blockchain-based encryption, it prevents any unauthorized access and information manipulation.
- *Network reliability*: The network layer ensures efficient data transfer with lower latency, especially for real-time IoT applications like autonomous vehicles or industrial monitoring.

3. *IoT-Blockchain layer* (*data processing With Blockchain integration*): The IoT-Blockchain layer integrates Blockchain technology with IoT data processing to enable transparent, tamper-proof, and automated operations.

- *Blockchain*: All of the data gathered is encrypted and safely stored on the Blockchain, which functions as a decentralized ledger. Each data point is saved as a "block" to create a chain once it has been validated. This guarantees data storage integrity, accountability, and transparency.
- *Smart contracts*: These are written directly into lines of code and function independently. Smart contracts use the information they collect to automate processes. For example, a smart contract might automatically start a supply chain payment when the package arrives. Once a shipment is delivered, the contract may automatically initiate payment.
- *Data analytics*: Advanced analytics may analyze the data due to Blockchain integration. Predictions, warnings, and IoT operation optimization all profit from the use of analytics tools, which may identify trends and anomalies in the data. The Blockchain provides sure that this data is trustworthy for decision-making and isn't tampered.

4. *Application layer* (*services and use cases*): After the data has been processed by the other layers, it is the topmost layer and primarily offers services to end users. The data and insights generated by this layer are localized into concrete applications across various industries.

- *Healthcare*: Signs like blood pressure, heart rate, and oxygen saturation can be tracked with IoT sensors. Blockchain facilitates the safe and accurate exchange of medical records between various providers, and smart contracts enable a number of tasks, such as scheduling an appointment or filing a claim.

- *Retail*: IoT components, including cameras and sensors, can be used to manage the supply chain, keep an eye on inventory, and comprehend consumer behavior. Additionally, smart contracts can be utilized to verify the provenance of goods, stop counterfeiting, and streamline payments in modern retail establishments.
- *Smart cars*: The integration of Blockchain technology and the Internet of Things (IoT) in smart cars enables real-time management of vehicle information, including fuel level, engine condition, and location. Vehicles and traffic control systems can communicate securely and transparently thanks to Blockchain technology. Examples of smart contract applications include tolls, parking, insurance claims, and payments from IoT devices that rely on data-driven handling.

Blockchain technology handles the data generated by IoT devices in each of these applications so that each action is recorded on an unchangeable, transparent ledger. For industries that depend on automation and real-time data processing, this increases efficiency and trust. This architecture ensures safe, effective, and transparent systems across a range of industries by combining the advantages of Blockchain and IoT technologies.

2.2.4 Key features of Blockchain for IoT

1. *Decentralization*: There isn't a single point of failure because no central authority is used.
2. *Immutability*: IoT guarantees the security and immutability of IoT data.
3. *Scalability*: Supports lightweight Blockchain solutions or sharding for IoT networks.
4. *Interoperability*: Makes it easier for diverse IoT devices to communicate with one another.
5. *Automation*: Some tasks can be completed without human intervention thanks to smart contracts.

Table 2.1 enumerates the potentials of Blockchain technology to transform IoT systems while highlighting approach contributions and future directions of research [1-13].

2.3 BENEFITS OF BLOCKCHAIN OF THINGS

2.3.1 Enhanced security and privacy

The enhancement of security and privacy in IoT ecosystems is the primary advantage that BCoT . Centralized servers have been used in traditional IoT networks. They are typically vulnerable to data breaches, cyberattacks, and

Table 2.1 Literature review

Reference	*Methodology / Approach Used*	*Key Findings / Contributions*
[1]	Comprehensive review of Blockchain technologies for IoT; classification of research issues such as scalability, privacy, and interoperability.	Described open problems of IoT and Blockchain integration, presented research gap in lightweight consensus algorithms and security models.
[2]	A survey on Blockchain-enabled Internet of Things security: architecture, consensus, and routing; classified based on the security feature and type of Blockchain.	Blockchain boost up IoT security via decentralization and transparency; Future work in optimizing energy and storage efficiency, security of infrastructure against robust attack can further improve the proposed model.
[3]	Blockchain-based IoT systems review; examined current security models and issues.	Illustrated how Blockchain could contribute to better managing trust in IoT; identified scalability and energy issues for practical deployment.
[4]	Literature review and theoretical framework for Blockchain in Edge of Things (EoT) using edge computing.	Suggested smart city edge computing applications with Blockchain; discussed challenges related to latency and security in Edge computing.
[5]	Review of literature on Blockchain applications and systems for IoT applications; discussed systems architecture, consensus mechanisms, and use cases.	Found Blockchain useful for securing transactions; recommended hybridization to improve performance against resource constraints.
[6]	Proposed a theoretical architecture for simulating Blockchain-based IoT ecosystems utilizing cloud computing	Generated a simulation environment to test Blockchain–IoT systems and assess scalability under changing workloads and verified the model interoperability with existing simulation environments.
[7]	Conceived a privacy-preserving IoT framework leveraging a combination of Blockchain and Privacy Set Intersection (PSI).	Permitted secure sharing of data with diminished privacy leakage while improving computational efficiency when compared to traditional schemes.
[8]	Developing and testing on IoT devices a lightweight Blockchain executable.	Illustrated the feasibility of Blockchain execution on constrained IoT hardware while revealing performance-territories regarding throughput and latencies.
[9]	Architectural modeling using principles of software engineering for IoT-enabled Blockchain.	Provided an architectural framework for a scalable IoT-Blockchain; increased modularity and security assurances.

(continued)

Table 2.1 (Cont.)

Reference	*Methodology / Approach Used*	*Key Findings / Contributions*
[10]	An examination of how Blockchain may be employed in the IoT ecosystem included consideration for security, scalability, and consensus protocols	Summarized the main Blockchain benefits for the IoT domain, along with the issues related to energy consumption and interoperability.
[11]	A survey was conducted of the integration of Blockchain and IoT, including a discussion of consensus protocols, security, and privacy.	A research roadmap was proposed for lightweight Blockchain protocols, and the importance of phenomenologically efficient key management and mechanisms to achieve trust was discussed.
[12]	A comprehensive survey of Blockchain applications within the IoT space focused on examples of use across domain contexts.	The benefits of Blockchain were highlighted in discussion of data integrity in IoT, and automation; hybrid approaches including both fog and edge computing in conjunction with Blockchain were suggested.
[13]	A systematic literature review (SLR) was conducted to analyze existing Blockchain-centric solutions and their potential for IoT.	Trends in design, and a gap in optimization of consensus protocols were described; design trends and future developments highlighted the necessity for standardization and interoperability.

single-point failure. By distributing data among several nodes and utilizing a decentralized architecture, Blockchain does eliminate these vulnerabilities and lowers the likelihood of unauthorized access. In addition, cryptographic hashing and encryption methods improve security even more while protecting user privacy.

2.3.2 Data integrity and transparency

Once written, data is unchangeable and impenetrable due to the immutable Blockchain ledger. When data integrity issues are highly relevant, this can be one of the pillars of IoT applications. Real-time data verification and traceability are made possible by BCoT, which is crucial for the financial services, healthcare, and supply chain management industries. In turn, this reduces the possibility of fraud, increases transparency, and fosters trust.

2.3.3 Decentralized control

Decentralized control over IoT devices has been made possible by the BCoT, enabling autonomous operations without the need for a central authority.

Smart contracts, which are code-based self-executable agreements, can be implemented on Blockchains to control how IoT devices interact with one another. This will increase efficiency and lower operating costs by automating processes like device authentication, data sharing, and payment processing.

2.3.4 Improved scalability

For IoT applications with millions of connected devices, a Blockchain is a distributed architecture that can accommodate up to a large number of nodes, making it extremely scalable. The inherent bottlenecks in centralized control systems cause scaling problems for the majority of IoT systems. On the other hand, P2P interactions that improve IoT network scalability can be facilitated by BCoT.

2.3.5 Secure microtransactions

IoT devices can conduct secure transactions and automate micropayments thanks to native integration of cryptocurrencies and tokens within the Blockchain. Devices must frequently carry out extremely low-value transactions without human intervention for applications like energy trading, driverless cars, and smart grids.

2.4 RESEARCH CHALLENGES IN BLOCKCHAIN OF THINGS

2.4.1 Energy consumption

It is also well known that Blockchain technology, specifically consensus mechanisms based on Proof of Work, consumes a significant amount of energy. IoT devices frequently fall short of the computational demands of Blockchain networks due to their frequently limited battery life and general computing power[5]. Because of this, creating consensus algorithms that use less energy is essential to BCoT's success.

2.4.2 Scalability and throughput

Blockchain still has a lower transaction throughput than traditional databases, despite its promise of scalability benefits. For instance, Bitcoin only completes about seven transactions every second, which is insufficient for widespread IoT applications. More research is required to develop more scalable Blockchain architectures. If the number of devices in IoT ecosystems is very high, sharding or Layer 2 scaling solutions are required.

2.4.3 Interoperability

The IoT ecosystem is expanded to include a variety of devices from various manufacturers, each with its own unique protocols and standards. In order for these devices to function well on a Blockchain network, interoperability issues between them must be resolved. Establishing guidelines and procedures for information exchange and interaction between various Blockchain networks and IoT devices is crucial.

2.4.4 Latency and real-time processing

Real-time data with low latency is necessary for the majority of IoT applications, including factory automation and self-driving cars. Given that Blockchain transactions have a lot of latency due to consensus processes and block verification, research is still desperately needed to lower latency without sacrificing security so that the BCoT can continue to support time-sensitive applications.

2.4.5 Privacy concerns

Although Blockchain offers tamper resistance and transparency, these features contrast with the need for privacy of applications. For example, certain IoT systems in use or finance would like to protect this sensitive data, even though it is against data protection laws to keep it on an immutable ledger. Secure multi-party computation and zero-knowledge proofs are two potential solutions to these issues.

2.4.6 Regulatory and legal challenges

Due to its decentralized structure, Blockchain presents difficulties with regard to legal and regulatory compliance. For instance, one might question how governments can enforce the law and who is accountable when something goes wrong in a decentralized network. To facilitate its widespread adoption, it is critical to create international regulatory standards for Blockchain applications in the Internet of Things.

2.5 FUTURE TRENDS IN BLOCKCHAIN OF THINGS

2.5.1 Development of energy-efficient consensus mechanisms

Research into novel consensus mechanisms that will reduce energy consumption without compromising security is one of the most exciting developments in BCoT. PoW in BCoT ecosystems is therefore being replaced by mechanisms such as Proof of Stake (PoS), Deligated Proof of Stake (DPoS), and Proof of Authority (PoA), which offer greater scalability and efficiency.

2.5.2 Integration with artificial intelligence (AI)

There is enormous potential in the combination of Blockchain, IoT, and AI. Blockchain ensures the security and integrity of the data, while AI algorithms can be used to analyze data from IoT devices. In domains like smart cities and healthcare, BCoT systems with integrated AI may enable autonomous operations, intelligent decision-making, and predictive maintenance.

2.5.3 Growth of Decentralized Autonomous Organizations (DAOs)

DAOs are essentially businesses that are controlled by smart contracts on a Blockchain. They will probably be crucial to the administration of IoT ecosystems. They manage resources transparently and without necessarily depending on trust, and they offer decentralized governance over BCoT applications that stakeholders can vote for. Additionally, this will probably pick up steam in the fields of supply chains, telecommunications, and energy management.

2.5.4 Adoption in smart cities

BCoT will be crucial to the development of smart cities, where IoT devices will be used to manage public services, utilities, transportation, and infrastructure. Blockchain improves data security and makes governance more transparent, allowing for safe and automated transactions between different systems in smart cities, such as transportation networks and energy grids.

2.5.5 Advancements in quantum-resistant cryptography

Today's Blockchain cryptographic algorithms are in some danger due to the discovery of quantum computing. To offer a safe route against upcoming quantum attacks, research into quantum-resistant cryptography for Blockchain is still ongoing. The creation of quantum-safe algorithms is still essential for long-term security in BCoT systems.

2.6 CONCLUSION

The emerging field of Blockchain of things has the potential to solve many of the security, scalability, and connectivity issues that IoT networks face. Combining the decentralized and unchangeable nature of Blockchain technology with the global connectivity of IoT can lead to significant advancements across industries. However, significant issues like latency, energy consumption, and regulatory compliance must be addressed before BCoT can be fully utilized.

Future research should focus on improving interoperability, developing Blockchain technologies that are both scalable and energy-efficient, and addressing privacy concerns in IoT applications. The BcoT has the potential to fundamentally alter a number of industries, including smart cities and healthcare, thanks to recent developments in artificial intelligence, consensus procedures, and cryptography.

REFERENCES

1. A. Ferrag, M. Derdour, M. Mukherjee, A. Derhab, L. Maglaras and H. Janicke, "Blockchain Technologies for the Internet of Things: Research Issues and Challenges," in *IEEE Internet of Things Journal*, vol. 6, no. 2, pp. 2188–2204, April 2019. https://doi.org/10.1109/JIOT.2018.2882794.
2. A. Ayub Khan, A. A. Laghari, Z. A. Shaikh, Z. Dacko-Pikiewicz and S. Kot, "Internet of Things (IoT) Security With Blockchain Technology: A State-of-the-Art Review," in *IEEE Access*, vol. 10, pp. 122679–122695, 2022. https://doi.org/10.1109/ACCESS.2022.3223370.
3. D. Commey, B. Mai, S. G. Hounsinou and G. V. Crosby, "Securing Blockchain-Based IoT Systems: A Review," in *IEEE Access*, vol. 12, pp. 98856–98881, 2024. https://doi.org/10.1109/ACCESS.2024.3428490.
4. T. R. Gadekallu et al., "Blockchain for Edge of Things: Applications, Opportunities, and Challenges," in *IEEE Internet of Things Journal*, vol. 9, no. 2, pp. 964–988, 15 Jan.15, 2022. https://doi.org/10.1109/JIOT.2021.3119639.
5. T. M. Fernández-Caramés and P. Fraga-Lamas, "A Review on the Use of Blockchain for the Internet of Things," in *IEEE Access*, vol. 6, pp. 32979–33001, 2018. https://doi.org/10.1109/ACCESS.2018.2842685.
6. A. Albshri, A. Alzubaidi, M. Alharby et al., "A Conceptual Architecture for Simulating Blockchain-based IoT Ecosystems," in *Journal of Cloud Computing*, vol. 12, p. 103, 2023. https://doi.org/10.1186/s13677-023-00481-z.
7. Q. Zhou, C. Lai, Q. Guo et al., "A Novel Privacy Protection Scheme for Internet of Things Based on Blockchain and Privacy Set Intersection Technique," in *Journal of Cloud Computing*, vol. 11, p. 93, 2022. https://doi.org/10.1186/s13677-022-00375-6.
8. N. Kullig, P. Lämmel and N. Tcholtchev, "Prototype Implementation and Evaluation of a Blockchain Component on IoT Devices," in *Procedia Computer Science*, vol. 175, pp. 379–386, 2020, ISSN 1877-0509. https://doi.org/10.1016/j.procs.2020.07.054.
9. W. Yánez, R. Bahsoon, Y. Zhang and R. Kazman, "Architecting Internet of Things Systems with Blockchain," in *ACM Transactions on Software Engineering and Methodology (TOSEM)*, vol. 30, pp. 1–46, 2021. https://doi.org/10.1145/3442412.

10. H. Hui, X. An, H. Wang, W. Ju, H. Yang, H. Gao and F. Lin, "Survey on Blockchain for Internet of Things," in *Journal of Internet Services and Information Security*, vol. 9, pp. 1–30, 2019. https://doi.org/10.1016/j.com com.2019.01.006.
11. H. Dai, Z. Zheng and Y. Zhang, "Blockchain for Internet of Things: A Survey," in *IEEE Internet of Things Journal*, vol. 6, pp. 8076–8094, 2019. https://doi.org/10.1109/JIOT.2019.2920987.
12. M. Ali, M. Vecchio, M. Pincheira, K. Dolui, F. Antonelli and M. Rehmani, "Applications of Blockchains in the Internet of Things: A Comprehensive Survey," in *IEEE Communications Surveys & Tutorials*, vol. 21, pp. 1676–1717, 2019. https://doi.org/10.1109/COMST.2018.2886932.
13. S. Lo, Y. Liu, S. Chia, X. Xu, Q. Lu, L. Zhu and H. Ning, "Analysis of Blockchain Solutions for IoT: A Systematic Literature Review," in *IEEE Access*, vol. 7, pp. 58822–58835, 2019. https://doi.org/10.1109/ACC ESS.2019.2914675.

Chapter 3

Advancements in Blockchain-based privacy-preserving framework for IoT

Keerthana N.V., M. Parimala Devi, R. S. Shudapreyaa, Prema P., and Madhumitha Kulandaivel

3.1 INTRODUCTION

The fusion of the Internet of Things (IoT) has reinvented diverse sectors, empowering uninterrupted connection among devices, users, and systems. IoT includes an enormous system of interconnected systems that gather, distribute, and inspect data in day-to-day applications, promoting effectiveness and creativity (Singh et al. 2023). Despite its advantages, privacy issues arise because of the large data size produced and such data sent to diverse platforms. Confidential data, varying from individual data to essential functional pieces of information, is frequently at risk of illegal access, overuse, or violations. Simultaneously, blockchain technology has evolved as a pioneering resolution for guaranteeing data protection, visibility, and consistency. Employing a distributed register system, Blockchain removes dependencies on individual authority, therefore improving the strength and consistency of data handling. Blockchain's intrinsic characteristics, like permanence and cryptographic security, make it a perfect candidate for handling IoT's confidentiality issues (Ahmed and Alabi 2024). The significance of Blockchain privacy in IoT can't be exaggerated. As fleets of IoT grow larger and more complex, preserving consumer trust and data privacy becomes paramount.

Technical advancements related to Blockchain, including interactive proof systems, secret smart contracts, and distributed persona management, offer creative approaches to preserving privacy in an IoT setting (Shen et al. 2023). This development not only solves the problem of privacy but also enables secure and flexible IoT applications, paving the way for a more secure digital future. In addition, the combination of Blockchain and IoT increases monitoring and accountability. Day-to-day data transactions among IoT devices can easily be tampered with and modified.

Blockchain provides the possibility of assuring not only that each data transaction is captured and cannot be tampered with back in time, but also reducing the risks and ensuring compliance with data management best practices. Its feature is invaluable, since such spheres as healthcare, smart

DOI: 10.1201/9781003584728-3

cities, and supply chain management rely on the integrity of information (Padma and Ramaiah 2024). Blockchain continues its evolution, whereas IoT growth is estimated to experience a significant uptick. Revolutions in progress, like suitable Blockchain formations along with adjustable synergism arrangements, are certain to conquer pretty much every contemporary deficiency, including extraordinary energy drain coupled with lag time.

These improvements will further help position Blockchain as an infrastructure for addressing confidentiality issues of IoT, together with security imperatives, thus driving toward a situation in which interconnected systems operate within a trustworthy, safe environment (Goyat et al. 2023). Among constraints of confidentiality—perhaps ultimate?—of IoT, we should name a highly concentrated world that has been built around traditional ways of handling data. Single servers have just a single point of failure, which represents an exciting target for those who take part in cyberactivity—a combination server system may face the same predicament; as a result, they become very attractive 'prey'.

One of the features of Blockchain that makes it hard for intruders to compromise the system is its decentralized architecture, where data is stored across multiple nodes (Zhang et al. 2023). However, this distribution does not bestow just security but also empowers consumers to take control over their data in a good way, which guarantees that personal data will be accessed or shared only with explicit consent. Additionally, privacy solutions based on Blockchain can effectively address issues related to system authentication as well as identity and access management in IoT environments (users) (Goyat et al. 2020; Akpakwu, Hancke, and Abu-Mahfouz 2020). Commonly used authentication systems often rely on centralized authorities who may sometimes be compromised (security breached) (Sai et al. 2023). Blockchain enables the utilization of distributed identity systems, whereby devices and users can securely prove their identities using cryptographic protocols (Hussain et al. 2021). Consequently, the need for any trustworthy external applications decreases, whereas trust among IoT residents enhances.

One function performed by Blockchain within IoT networks is enabling data monetization and exchange without sacrificing privacy at the same time (privacy preserving). Specifically speaking, the IoT generates huge volumes big volume confederated unrefined input, valuable purpose research, and analysis business (Shen et al. 2023). The use of a secure and transparent structure supported by Blockchain paves the way for shared rights to intermediaries who would transact data judiciously into smart contracts, demanding certain conditions so that entitlements are protected from exposure, thus creating value-added economic activities while ensuring protection of digital information, which does not conflict with the requirements General Data Protection Regulation (GDPR) or other regulations.

Although it is advantageous, there are some disadvantages to using the Blockchain framework. Flexibility and power efficiency remain important issues, as the usual Blockchain networks frequently face hardships in dealing with large bundles of transactions typical of IoT systems (Rahmadika et al. 2022; Khan et al. 2016). Yet continuing research in regions like Layer 2 Solutions, partitioning, and power-productive accord mechanisms, like Proof of Stake (PoS), is offering the path for more down-to-earth applications. These innovations are crucial to confirming that Blockchain can meet the exigent needs of large-scale IoT implementations (Padma and Ramaiah 2024). It can be concluded that the combination of Blockchain and IoT is a new approach to security and privacy establishment in networked systems. In such a context, Blockchain, with its features of distributed, transparent, and secure data management, is the best fit for IoT-connected devices (Goyat et al. 2020). And as we evolve as technology innovators, the ability to meld that with IoT will open up levels of trust, capacity, and advancement that we have never seen before, that will define the future of communities and companies around the world.

There are a wide range of technologies that are applied in context-sensitive IoT systems to gather, understand, and apply the context information for making reasonable decisions. The first step of this process is to capture the information or to collect the data from the sensors, IoT devices, and edge computing platforms. For a good understanding and taking action within the IoT environment, modeling methods, such as ontologies and context-aware middleware frameworks are used to integrate the data collected (Rahmadika et al. 2022). AI and machine learning (ML) give us a higher ability to analyze and understand this data and enable computers to make context-sensitive decisions (Schürholz, Kubler, and Zaslavsky 2020). Transmission of the contextual information is also enabled due to data exchange protocols, like message queuing telemetry transport (MQTT), which was utilized in Singh et al. (2023) and constrained application protocol (CoAP) (Zhang et al. 2023), especially in distributed and mobile scenarios. To gather and process context data in real time, cloud and edge computing are needed to enhance system efficiency and mobility.

3.2 BLOCKCHAIN-BASED PRIVACY-PRESERVING FRAMEWORK

3.2.1 Fundamentals

Blockchain aggregation with the IoT combines Blockchain's distributed, immutable, and transparent nature with the scale and the interconnection scope of IoT networks (Goyat et al. 2023). This partnership helps to solve some of the main IoT limitations, particularly when it comes to confidentiality and data security. IoT generates much data, and Blockchain guarantees secure data exchange between devices using the decentralized register system. This eliminates the need for centralized entities, mitigates concerns about data tampering, and provides a trustless environment where system communication is authenticated and validated without the need for a middleman.

3.2.1.1 Smart contracts

Smart contracts contain pre-coded agreements that act independently. They play an important part in IoT surroundings by automating communications among systems and ensuring safe and effective data sharing. For instance, a smart agreement can empower an IoT-empowered house to automatically pay electric bills based on day-to-day electricity usage, lowering human intervention and faults. The mathematical formula for smart contracts is as follows: ,

$$\text{Execute}\left(B_i\right) = \begin{cases} 1, \textit{if } D_i \textit{ is satisfied} \\ 0, \textit{ otherwise} \end{cases} \tag{3.1}$$

where B_i is the action i to be performed.
D_i is the condition i that should be approved for B_i to be performed.
$\text{Execute}\left(B_i\right)$ describes whether B_i action is performed (1 indicates success and 0 indicates failure).

3.2.1.2 Decentralized mechanisms

IoT devices usually depend on concentrated networks, which are vulnerable to single-point faults and security issues. Blockchain removes these issues by scattering data among a network of nodes. The distributed register assures that each system's operations are safely documented and unchangeable, preserving against spoofing and unauthorized manipulations.

3.2.1.3 Consensus algorithms

The agreement algorithm allows Blockchain network nodes to agree on the legitimacy of the communications. IoT applications frequently apply flimsy algorithms to ensure their suitability with systems that have restricted computational efficiency. Well-known agreement algorithms are incorporated. A general formulation for consensus possibility is represented in (3.2):

$$P'(\text{Consensus}) = 1 - P'(\text{Fault}) \tag{3.2}$$

where $P'(\text{Consensus})$ is the possibility of achieving consensus.
$P'(\text{Fault})$: Possibility of failure to achieve consensus, usually shaped by the number of spiteful nodes.

Proof of Work (PoW): Fortressed but computationally rigorous, compatible with situations where security is important. In PoW, consensus depends on the possibility of an individual node handling the cryptographic enigma initially. The possibility of a node j effectively extracting the consecutive Block is in (3.3),

$$P'_j = \frac{Hash\ Rate_j}{Total\ Hash\ Rate} \tag{3.3}$$

Proof of Stake **(PoS):** Power-effective and highly compatible for IoT ecosystems with restricted energy demands. In PoS, consensus relies on the ratio of the stake owned by the genuine nodes. The possibility of achieving consensus can be represented as in (3.4):

$$P'(Consensus) = \sum_{j \in H'} \frac{Stake_j}{Total\ Stake} \tag{3.4}$$

where H' indicates a set of genuine nodes
$Stake_j$ indicates the stake owned by the j node.

Practical Byzantine Fault Tolerance (PBFT): Refined for minimal delay IoT networks, it offers quick and trustworthy accord. For PBFT, the possibility of consensus is represented in (3.5) and is shaped by the number of spiteful nodes.

$$P'(\text{Consensus}) = 1 - \left(\frac{k}{n'}\right)^f \tag{3.5}$$

where k indicates the number of spiteful nodes.
n' indicates the number of nodes.
f indicates the number of iterations for consensus.

3.2.2 Key Features

Blockchain's inherent features significantly enhance IoT privacy and security.

3.2.2.1 Transparency

Blockchain transactions are logged in a public and partially public register, assuring responsibility and traceability. This integrity is indispensable in applications like supply chain IoT, where consumers need transparency about product quality. By offering a traceable path of communications, Blockchain promotes trust among candidates and prevent the issues of forgery.

3.2.2.2 Pseudonymity

Blockchain balances transparency with confidentiality by linking transactions to pseudonymous addresses instead of personal identities. This assures that although system communications persist noticeably, confidential user information is secured. For example, an IoT-empowered portable can transfer health information to a Blockchain without revealing the consumer's identity.

3.2.2.3 Immutability

Blockchain's unchangeable nature ensures that data once entered can't be manipulated or removed. This is crucial for IoT systems that need tamper-proofing records and audits, like legal and regulatory conformity and forensic inquiry. Unchangeability improves data consistency and assures that all consumers have a steady view of the device's condition.

3.2.2.4 Data encryption

Blockchain retains advanced encryption strategies to safeguard IoT data. The usual methods include elliptic curve cryptography (ECC) and hash functions like SHA-256. These ensure data consistency and privacy are preserved across the network. The formula in (3.6) generates a unique fixed-length output for any given input, making it computationally infeasible to reverse-engineer or modify the original data. The general ECC is represented in (3.7).

$$h' = H'(y) \tag{3.6}$$

where y indicates the input data.
$H'(y)$ indicates a hash function.
h indicates the fixed-length output.

$$a^2 = b^3 + yb + c \; mod \; p' \tag{3.7}$$

where y and c are the coefficients of the curve.

p is the prime number defining the finite field $F_{p'}$.

a and b are the points on the curve.

3.2.3 Architecture and models

Blockchain architectures serve multiple IoT requirements, each providing exceptional advantages and compromises.

3.2.3.1 *Public Blockchain*

Public Blockchains are open to all candidates and assure greater transparency. They are steady for insensitive IoT applications, like environmental observations and public management applications. Even though their higher power usage and lag, because of consensus mechanisms such as PoW, pose limitations for day-to-day IoT operations. The lag in public Blockchain emerges from various factors like Block propagation delay, consensus mechanism delays, and network congestion. A general mathematical formula for latency L' in a public Blockchain can be represented as in (3.8):

$$L'_{public} = T'_a + T'_b + T'_c \tag{3.8}$$

where L' is the total latency in the Blockchain system.

T'_a is the Block creation time.

T'_b is the Block propagation time.

T'_c is the consensus time.

3.2.3.2 *Private Blockchain*

Private Blockchains are restricted to specific participants, offering better control, scalability, and faster transaction times. They are suitable for industrial IoT applications with critical data sensitivity and access control. However, reduced decentralization may compromise trust. Private Blockchain is limited to a particular candidate, providing improved control, flexibility, and speed of transaction period. They are appropriate for industrial IoT applications where data integrity and access control are crucial. Yet, lessened distribution may trade off trust. The total latency is represented as in (3.9):

$$L'_{\text{private}} = T'_a + T'_b + T'_c \tag{3.9}$$

where L' is the total latency in the Blockchain system.

T'_a is the Block creation time.

T'_b is the Block propagation time.
T'_c is the consensus time.

3.2.3.3 *Consortium Blockchain*

Consortium Blockchain is controlled by a cluster of organizations, striking an equilibrium between public and private models. They are especially helpful in sectors such as healthcare and finance, where several consumers need shared access to information. Governance intricacy and cooperation among candidates are limitations that should be addressed. The total latency is represented as in (3.10):

$$L'_{\text{consortium}} = T'_a + T'_b + T'_c + T'_d \tag{3.10}$$

where L' is the total latency in the Blockchain system.
T''_a is the Block creation time.
T'_b is the Block propagation time.
T'_c is the consensus time.
T'_d is the governance delay.

3.2.4 Real-world examples: Supply chain management

Supply chain management is among the foremost transformative areas where Blockchain aggregated with IoT has been described important potential. The fusion of these technologies addresses usual supply chain limitations, like vagueness, lack of efficiency, forgery, and hardship in monitoring product sources. By utilizing Blockchain and IoT, supply chain systems can become highly transparent, effective, and trustworthy.

3.2.4.1 *Core elements of Blockchain in supply chain*

Origin tracking: IoT systems like radio frequency identification (RFID) tags, global positioning system (GPS) trackers, and temperature detectors produce real-time information regarding wares at several phases of the supply chain (Hussain et al. 2021). This information is logged unchangeably on a Blockchain, offering a tamper-proof inspection. For instance, the parcel of ephemeral goods, such as products of the sea, can have temperature detectors that monitor the status during the transfer (Kaur et al. 2022).

End-to-end transparency: Blockchain assures that every transaction and change in the supply chain is logged on a decentralized register. This removes the differences and assures noticeability for all candidates, including producers, suppliers, and consumers.

Smart Accords for Automation: Smart agreements automate operations like the release of payment, verification of orders, and compliance audits. For example, once the parcel is shipped and verified by the IoT detectors, a smart agreement can automatically initiate payment to the distributor. Numerically, a smart contract for shipment validation and payment can be described as in (3.11):

$$P' = \begin{cases} A' \ if \ D' \geq D'_{min} \ and \ T' \leq T'_{max} \\ 0 \ otherwise \end{cases} \tag{3.11}$$

where P' indicates payment amount.
D' indicates the distance covered by the shipment.
D'_{min} indicates the minimum required distance.
T' indicates the transit time.
T'_{max} is the maximum allowable transit time.

3.3 ADVANCEMENTS IN BLOCKCHAIN FOR IOT FRAMEWORK

3.3.1 Emerging technologies

Integrating the technologies of Blockchain with innovations such as edge computing, AI, and homomorphic encryption will enhance the current privacy and security standards in the framework fields of the IoT platform.

3.3.1.1 *Edge computing*

Edge computing is a promising tool to overcome these limitations and does not rely on centralization, and each node in edge computing is integrated with Blockchain, serving as a validator responsible for updating and processing transactions. Sensitive data like location and data behaviors are protected well in real-time in edge computing. The time taken to process the data is represented in (3.12):

$$T_{\text{edge}} = T_{\text{in}} + T_{\text{compute}} + T_{\text{op}} \tag{3.12}$$

where T_{in} is the time taken to receive the data from the IoT devices.
T_{compute} is the time taken for data processing.
T_{op} is the time taken to send the processed data to the respective destination.

3.3.1.2 AI Rules the World

Integrating Blockchain and IoT with AI enables automated decision-making, which plays a crucial role in many applications. Blockchain is highly useful in keeping the recorded events unchanged and uses AI to keep suspicious activities in check and to alert the system when security breaches occur. AI is also used to increase resource optimization. Estimating the probability of anomaly detection and fraud detection is represented in (3.13) and (3.14), respectively.

$$p_{Anom} = \frac{\text{No. of anomalous samples detected}}{\text{Total samples analyzed}} \quad (3.13)$$

where p_{Anom} is the anomaly probability.

$$e_{Fraud} = \frac{\text{No.of fraudulent transactions detected}}{\text{Total fraudulent transactions}} \quad (3.14)$$

where e_{Fraud} is the detected fraudulent transaction.

3.3.1.3 Homomorphic cryptosystem

To handle sensitive data with care, privacy is vital. Homomorphic encryption is used to provide data security to highly sensitive records by encrypting data without the requirement for decrypting. This gives more protection of data, eliminating unwanted threads. It is highly useful when there is a need for high secrecy of sensitive data. The encryption and decryption processes are represented in equations (3.15) and (3.16), respectively.

$$C = g^m.r^n \bmod n^2 \quad (3.15)$$

$$m = \text{Decrypt}(C) \quad (3.16)$$

where C is ciphertext.
g is the generator.
r is the random number.
m is the plain text.
n is the modulus.

By offering strong privacy and protection to the data, data processing, and enhancing the current limitations, the usage of Blockchain with emerging technologies improves the IoT frameworks and qualifies them for more scalable and practical applications.

3.3.2 Scalability

To adopt the IoT frameworks in the Blockchain, scalability plays a crucial role. By ensuring scalability, millions of devices can simultaneously transact without facing any issues. The following new advancements have been proposed a solution to overcome the limitations and challenges of traditional architecture.

3.3.2.1 Sharding

Sharding segments the Blockchain network into finer segments, known as shards, where each shard handles its subset of transactions. This method permits various shards to process simultaneously, significantly raising transaction capacity. For IoT applications, sharding is especially helpful as it distributes the work burden among the networks, lowering the constraints. For instance, in a smart city, various shards can manage individual data channels, such as public safety and transportation, ensuring effective processes as presented in (3.17):

$$t_{\text{shard}} = \frac{n}{s} \tag{3.17}$$

where n denotes the total transactions made.
s denotes the number of shards used.

3.3.2.2 Off-chain solutions

Off-chain solutions, such as lightning network and off-chain transaction hubs, unload transactions from the primary Blockchain. These solutions include making a secondary layer where transactions are managed and only the final integrated outcomes are logged on-chain. This importantly lessens overcrowding and enhances the speed and thrift of Blockchain networks. For example, an IoT-empowered smart factory can manage microtransactions for equipment processes off-chain, logging only abstracts on the Blockchain (3.18):

$$R_{eff} = \frac{1}{n} \tag{3.18}$$

3.3.2.3 Layer-2 technologies

Layer-2 solutions, like rollups (optimistic rollups and zk-rollups), bundle various transactions into an independent condensed transaction before offering it to the Blockchain. This lessens storage and computational needs, empowering IoT systems to communicate with Blockchain networks without overpowering them. For instance, in independent vehicle networks, rollups can integrate various vehicle communications and offer them as individual logs to the Blockchain, assuring effective data handling. The compression time is calculated in (3.3) as follows:

$$t_{\text{comp}} = \text{Compress}(t1, t2, \ldots, tN) \tag{3.19}$$

These flexibility innovations ensure that Blockchain technology can manage large data bundles produced by IoT systems, making it realistic for applications in various sectors like agriculture, healthcare, and logistics.

3.3.3 Security enhancements

Safeguarding IoT devices is critical because of the susceptible nature of the information they manage. Blockchain improves security using distributed trust, but modern advancements additionally improve its capacity to preserve privacy and safeguard data.

Ring signatures: Ring signatures are incognito as they permit a transaction signer to combine their signature with a clique of potential signers. This assures the transaction is validated without exposing the actual signer (Wang et al. 2023). In IoT applications, like smart healthcare, ring signatures allow systems to exchange crucial medical information confidentially, safeguarding the privacy of patients while assuring data security.

Zero-knowledge proofs (ZKP): ZKP technologies, like zero-knowledge succinct non-interactive arguments of knowledge (zk-SNARKs), enable the individual body to prove the reality of an assertion without revealing the fundamental data. For example, an IoT-authorized home safety system takes care of validating a user's identity to avoid dismissal of responsibility without giving impressionable credentials. In modern IoT, ZKPs can guarantee secure device confirmation without exposing encryption solutions or maneuver-specific data.

Secure multi-party computation (SMPC): SMPC admits multiple parties to calculate a function over their inputs without disclosing the inputs themselves. When linked with Blockchain, SMPC allows secure cooperation between IoT instruments.

3.3.4 Interoperability and standards

Interoperability is an important factor to allow Blockchain in IoT to be widely accepted because it ensures no friction of ideas and cooperation from different devices and networks. Yet, new shoots dedicate labor to something that would establish some common guidelines and duties to promote unity.

Cross-chain communication: Cross-chain computing, such as Polkadot and Cosmos, would allow various Blockchain networks to communicate and exchange data securely. Cross-chain designs in IoT systems facilitate the consolidation of heterogeneous schemes and platforms. An illustrative use case could be a smart home ecology managing household-related user scenarios and multisided platform chatter between various Blockchain networks (e.g., Ethereum and Hyperledger), enabling coordinated operation and data sharing.

Open standards development: Entities such as the IEEE and ISO are busy on normative Blockchain-IoT platforms. Anti-interoperability is countered through "common dossier layouts", APIs and communication contracts as part of these guidelines. Standardization accelerates adoption by reducing complexity and enabling compatibility from maneuver to maneuver and between networks.

Middleware Solutions: Middleware platforms are intermediaries between Blockchain networks and IoT devices, which translate information formats and guarantee a match of the contract. These choices accelerate the adoption of Blockchain in traditional IoT environments and allow the application of those who are willing to embrace Blockchain without having to reinvent the wheel. For instance, middleware may also integrate Blockchain-enabled smart grids with older legacy energy management systems, thereby enabling a seamless transition. These developments, bringing interoperability and standardization, form an ecosystem where IoT devices, applications, and connectivity can interoperate not only more efficiently but also enable them to drive innovation.

3.4 CONTEXT-AWARE ENABLING TECHNOLOGIES

3.4.1 Sensing and data collection

Inspired by the advent of semantic communication, a semantic transmission model is proposed to report the sensing data of the physical world to the Metaverse. It is possible to transmit only a minimum amount of data by allowing semantic encoding of the sensing data by employing the semantic bases. This procedure maintains performance and, not invoking data recovery, achieves the sensing objective. The contest theory is used by the authors in this paper to propose an incentive mechanism in the auction that can motivate users to contribute their content more frequently to help the service quality of the Metaverse (Wang et al. 2023). Here, the structured approach, in the spirit of developing the Semantic Web, was influenced, proposing a semantic transmission mode (STM) as a paradigm for transmitting sensing data to the Metaverse (Huang et al. 2024). Setting of semantic representation to the sensory data is performed by the system by using the semantic bases, which reduces the size of the data to be transmitted. By doing so, performance is retained, as well as the sensing goal can be achieved without data recovery.

Some of the location-based services utilizes technologies like GPS, Bluetooth beacons, and WiFi. A context-aware adaptive data cube (DC) called the C2A-DC enhances data analysis by adopting to contextual information (Temenos et al. 2024). It was created to track environmental variables to lessen the consequences of climate change. It integrates data harmonization, deep and ML algorithms for categorization, DC generation, and the computation of remote sensing (RS) activities. Because of the framework's context awareness, environmental monitoring tasks can be customized to meet the needs of stakeholders. Additionally, it removes the requirement for end users to perform extra processing steps. Two real-world scenarios are used to demonstrate the C2A-DC framework (Temenos et al. 2024): convolutional neural network (CNN) architectures for cloud and shadow detection during flood events, and RS products for fire-related deforestation monitoring.

In deep learning-based trackers, the combination of search region and template attributes is essential. Transformer-based trackers employ attention processes, whereas Siamese-based trackers frequently employ cross-correlation procedures. There is a lack of comprehension because existing trackers employ the same structure for both regression and classification heads. A feature enhancement-fusion network (FEFN) (Zhang, He, et al. 2024), which combines the benefits of transformer-based and Siamese-based trackers, is suggested as a solution to these problems. This tracker successfully Blocks distractions and creates worldwide connections. Additionally, a new decoupled head is suggested, which improves the context-aware tracker's accuracy.

Some of the user activity involves tracking equipment like wearables, accelerometers, and gyroscopes. Significant progress has been made in many areas as a result of the quick development of technology, but two of the most revolutionary developments are edge computing and the IoT. Instead of depending on a centralized data-processing warehouse, edge computing involves processing data close to the point of data production. Reduced latency, increased bandwidth efficiency, and higher security are just a few advantages of this strategy. Conversely, the IoT is a collection of actual objects that are linked together via the Internet so that they may gather, exchange, and analyze data. To fully utilize the promise of both technologies, edge computing and IoT must work in tandem.

3.4.2 Modeling and representation

For structuring and representing context data for meaningful interpretation, diverse parties must exchange data and information due to the energy sector's increasing interoperability requirements, which frequently involve major context disparities. To improve interoperability, Jenevain et al. (2024) suggest an addition to the open energy ontology (OEO) that focuses on context. It explores how context shapes comprehension and suggests inferring techniques and strategies.

Now, by using contextual data, they increase the relevance and dependability of actions for both consumers and the IoT ecosystem. Context adoption is significantly advancing applications that are time- and safety-critical, such as autonomous driving. Although there is a significant increase in the sources of context information to support this trend, their quality and performance are inconsistent, and their features—such as their presentation and interfaces—vary greatly. Untangling these intricacies and providing relevant context information in response to each context request is costly and not scalable in a vast IoT ecosystem. Li et al. (2024) examines and suggests a new method to give context management systems (CMS) an adaptive context caching so that context data can be reused and repurposed for quick concurrent processing.

Some of the fusion approaches are as follows: the foundation for building visual question answering (VQA) models has been the question model. To model multi-level inquiries, current question models try to take advantage of word context to extract multi-level notions, but have a lot of flaws, though. Context-aware multi-level question embedding fusion (CMQEF) (Hu et al. 2024) is a revolutionary, low-complex, multi-level contextual question model that we propose to address these problems.

Building a human-centered intelligent driving system requires accurate vision-based driver attention assessment, which is a difficult undertaking because of the visual sensor's limitations. To jointly model the scene and the driver's state while driving, a data-driven hybrid architecture for context-aware driver attention estimation is developed (Li, Li, Zhang, et al. 2024). It is generally considered that a visual saliency map will draw attention to a particular location. A multi-hierarchy fusion network is suggested as a way to make use of this feature and efficiently extract saliency information from a scene image.

Particularly, in the context of social media and meta-universe applications, where communication transcends text to incorporate images, videos, and audio, sarcasm—a subtle and complicated type of expression—presents considerable detection issues. Emotion nuances identical to incongruence are the sarcastic element, which is usually not captured by text-based sarcasm detection methods. To address these challenges, a new multisource sarcasm detection approach (Hassan et al. 2024) that explicitly captures the discrepancy of emotions across different modalities—a crucial element that the classical approaches may sometimes neglect, besides handling multimodal data.

3.4.3 ML and AI in context awareness

ML/AI plays a role in context data processing for context data analysis and processing. Using NP-hard problems to properly process the context data, the authors propose an efficient context-aware service offloading mechanism in the framework of mobile cloud computing (MCC) (Hassan et al. 2024). The new ML model relies on task adaptation to select which tasks to offload. Considering mobile device time, cost, network, location, and CPU power to train data, the mobile charging vehicles (MCVs) scheduling algorithm predicts utilization rates of specific microservices for a feasible task scheduling scheme. The higher efficacy of the proposed system is the lower cost and energy consumption in offloading microservices to MCC. The microservice offloading method in MCC is effective in conserving energy and money. In the area of clustering, to significantly alleviate the burden of security operation centers (SOC), the authors propose a context-aware clustering mechanism targeting security alerts classification (CALLEE) (Liu et al. 2024) that leverages broad relationships between alerts to locate related ones. The approach succeeds in identifying relevant warnings by applying clustering, meta-paths, and a core concept model. Although the results may not be perfect, security analysts can inspect a few scores within the clusters manually by factorizing downward the number of clusters and the homogeneity within every cluster. The approach has been proven and is applicable and implementable in industrial settings.

The XGBoost ML algorithm is used for predicting therapy for hepatocellular carcinoma (HCC) in patients (Daye et al. 2024). According to the results, the algorithm can correctly forecast each of the board's eight therapy recommendations, which could help clinicians make decisions in situations where specialized knowledge is weak. In real-time, three ML models (Kolbinger et al. 2024) were trained for surgical phase recognition, and individual segmentation models were trained using DeepLabv3.

3.4.4 Communication protocols for mobility

For IoT applications, a smooth platform connection is essential, and work is being done on auto-configuration to reduce human intervention. The standardization group OneM2M exchanges data using MQTT, hypertext transfer protocol (HTTP), and CoAP. However, because of its shortcomings in resource directories and multicasting, MQTTc (Kim et al. 2019), it is not appropriate for service discovery. An architecture for an auto-configuration method utilizing MQTT publish/subscribe communications is shown in this

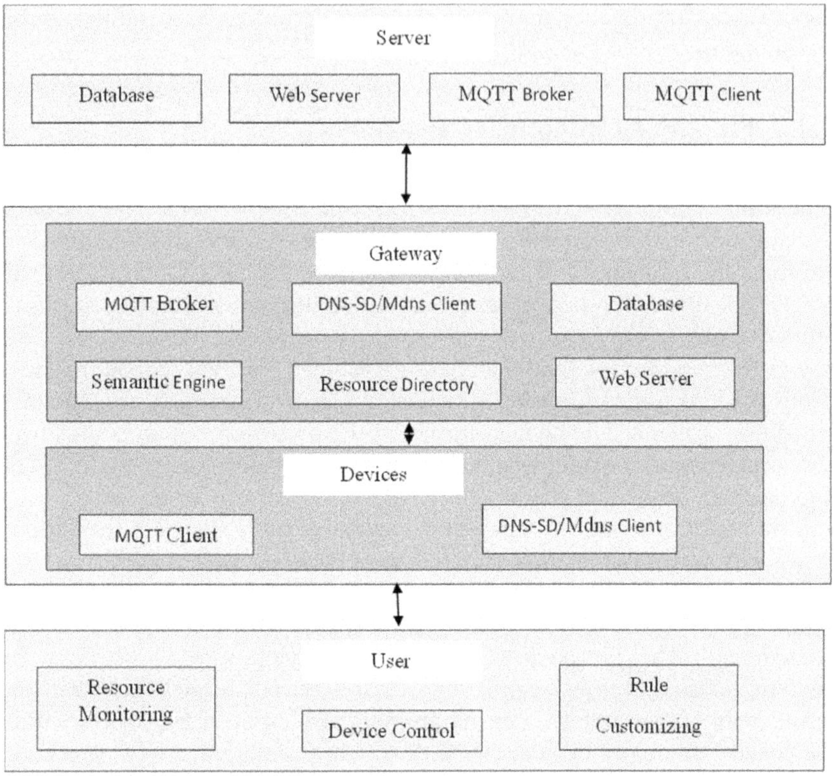

Figure 3.1 A context-aware MQTT communication protocols for mobility.

work. The system uses semantic Web technologies to identify context to communicate with gadgets. Through practical application, the viability of the suggested system is confirmed, demonstrating its ability to automatically configure MQTT-based devices (Kim et al. 2019) and offer context-aware services through a common interface as shown in Figure 3.1.

In networks of limited devices, the IoT has created a need for effective congestion control systems. One common communication protocol that facilitates lightweight interoperability is the CoAP (Bormann, Castellani, and Shelby 2012). It is essential to distinguish between different packet loss circumstances and determine the accurate round-trip time (RTT) of retransmitted message acknowledgment because the underlying CoAP specification is insensitive to diverse network conditions. A context-aware congestion control (CACC) method for lightweight CoAP/user datagram protocol-based IoT traffic is presented in this work. The CACC (Akpakwu, Hancke, and Abu-Mahfouz 2020) suggests strategies like the aging idea, lower bound

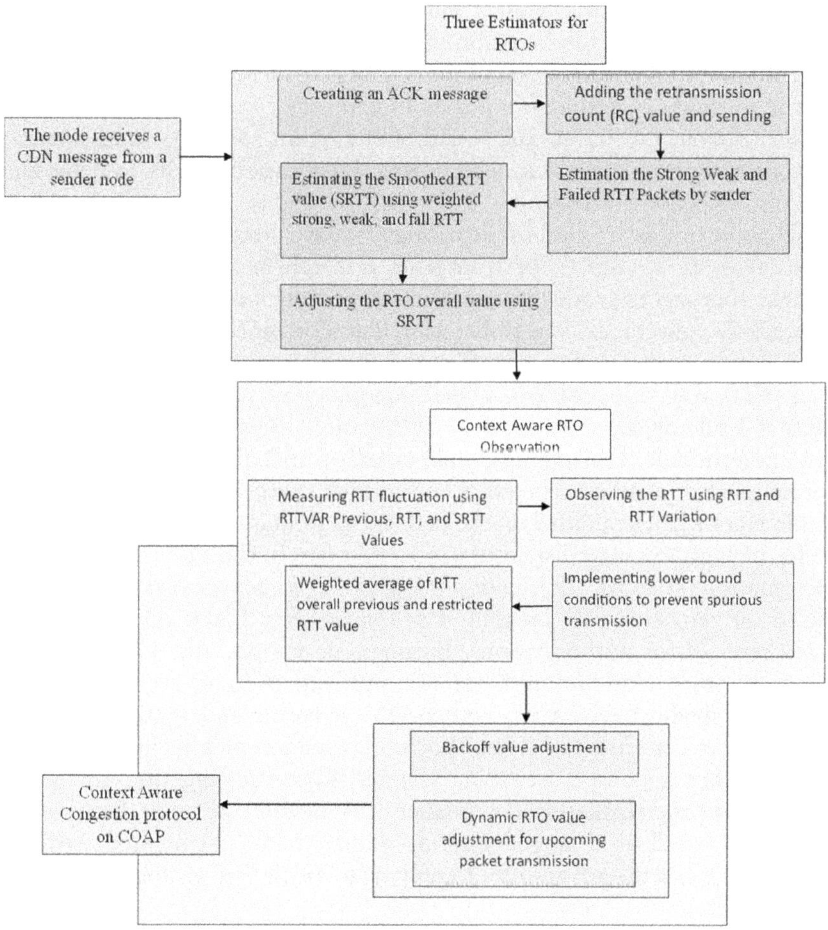

Figure 3.2 A context-aware CoAP Communication protocol for mobility.

recovery time objective (RTO) limiting approach, retransmission count-based smoothed round-trip-time observation, and retransmission timeout estimator. The suggested RTO estimators allow adaptive congestion control and pinpoint the precise network status using strong, weak, and failed RTT as shown in Figure 3.2.

3.4.5 Cloud and edge computing overview

ML, video comprehension, and the placement of video clips all depend on scene text recognition. However, the high model complexity and significant detection delay of current approaches make it difficult to identify irregular texts in photos. Using cloud and edge layer computing resources, a context-aware edge-cloud collaborative scene text recognition technique is suggested. Rapid classification of multi-category scene text is made possible by the method's introduction of a multi-category context awareness mechanism (Zhang, Liu, et al. 2024). Additionally, an adaptive multi-category detection method is created, which uses classification results to determine the best detection strategy.

3D point clouds are vital for depicting plant architecture, and plant phenotypic analysis is critical for transgenic research and breeding. However, current approaches produce incomplete 3D point clouds due to their inability to precisely collect local and global data. PlantFormer (Li, Li, Qi, et al. 2024), a revolutionary fine-grained point cloud completion technique, is suggested as a solution to this problem. This technique preserves local regional information while aggregating local–global features using an edge-convolution attention module. For fine upsample effects, a spatial context-aware transformer is also included. The model performed exceptionally well on several parameters when evaluated on the high-quality dataset PlantCom3D.

To manage access to dispersed medical devices in Cloud-based healthcare information systems, the paper suggests a context-aware architecture based on zero trust. By creating a scoring system based on cloud-native microservices for authorizations, logging, encryption, and authentication, the framework (Al-Hammuri, Gebali, and Kanan 2024) seeks to prevent or reduce medical errors. To evaluate the semantic and syntactic analysis of attributes contained in a healthcare information system in real time, it also develops a bond trust rating system. When creating the framework, user consent and regulatory compliance are taken into account. Because this method is based on a language model rather than a set of predetermined attributes, it has the advantage of applying to any language and is adaptable to all attributes.

In Cloud-based healthcare, this strategy is essential for creating safe tools for dispersed medical equipment. Cloud computing enables scalable context data processing, edge computing ensures real-time local processing with reduced latency, hybrid cloud-edge solutions balance efficiency, and use cases include smart cities and autonomous vehicles.

3.5 APPLICATIONS OF CONTEXT-AWARE ENABLING TECHNOLOGIES

Systems are being revolutionized by context-aware technologies, which use real-time data to provide individualized and intelligent answers. They maximize operations in smart cities, healthcare, smart homes, and automation, automation vehicles, and industrial IoT, as shown in Figure 3.3. These technologies are influencing innovation and advancement.

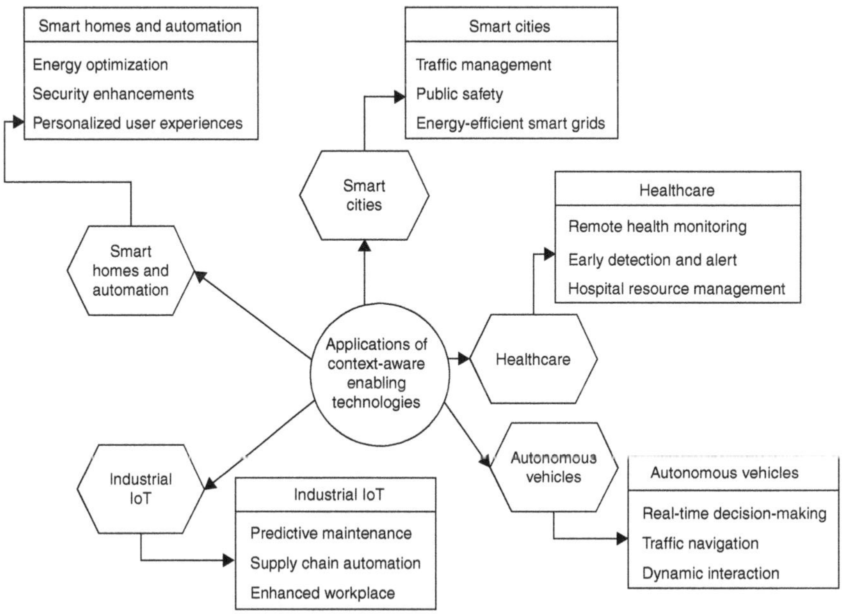

Figure 3.3 Applications of context-aware enabling technologies.

3.5.1 Smart cities

ICT-based solutions are discussed in a paper in a way that they may be used to overcome resource limits and global urbanization. The authors propose a smart city model that connects the main infrastructures, such as utilities, transportation, healthcare, and the environment, with the help of fog computing and context-aware computing (Reddy et al. 2020). The research points out a virtual machine management plan that uses reinforcement learning (RL) and genetic algorithms (GA) to lower the workload of fog nodes, thus reducing the power consumption and keeping good quality of service (QoS) levels.

ICT-based solutions are applicable in adverse resource conditions and urbanization all over the world. The system describes a smart city model that carries out the management of vital infrastructures such as utilities, transportation, healthcare, and the environment through the use of fog computing and context-aware computing (Schürholz, Kubler, and Zaslavsky 2020) to lessen the energy consumed and keep up with high-QoS obligations. Massive quantities of data serve as the backbone for smart cities aiming at improving social life, healthcare, and transport.

On the other hand, such data often contains some users' privacy; therefore, it is necessary to implement privacy protection before the data can be used for analysis and forecasting. To develop a privacy-aware data fusion and prediction method for smart city industrial contexts, Qi et al. (2020) resort to the traditional locality-sensitive hashing algorithm. Experiments conducted in the actual world demonstrate that the strategy outperforms other competing approaches in terms of prediction performance. The difficulty is in striking a balance between data availability and privacy, which makes accuracy and privacy difficult to achieve.

3.5.2 Healthcare

Digital twin (DT) is a promising technology in the field of digital and smart healthcare, transforming the way patients and healthcare professionals interact. DT is a virtual replica of a physical asset that reflects real-time data, enhancing healthcare operations. Elayan, Aloqaily, and Guizani (2021) propose an intelligent context-aware healthcare system using the DT framework, implementing an electro-cardiogram (ECG) heart rhythms classifier model. The model successfully predicted heart conditions with high accuracy in various algorithms. Integrating DT with healthcare would improve processes by bringing patients and professionals together in an intelligent, comprehensive, and scalable health ecosystem. Neural-network-based algorithms are better suited for ECG data than traditional ML algorithms.

Applications that are sensitive to latency can benefit from fog computing, which makes Cloud-based resources available to end users. However, scheduling jobs is difficult due to resource heterogeneity, dynamic nature, resource restrictions, and unpredictability. Numerous scheduling techniques that take contextual information into account have been put forth to address this. The comparison of current methods based on context-aware parameters, case studies, performance measurements, and assessment tools offers a thorough literature analysis of context-aware scheduling in fog computing (Islam, Kumar, and Hu 2021). Additionally, it offers context-aware parameter analysis, performance metrics, and taxonomy.

To handle medical emergencies, Ghosh et al. (2020) suggest a system known as cloud-fog based workflow for emergency service (CLAWER). Smartphones serve as edge devices, and they employ IoT sensors to gather health and mobility data. The fog device processes the data, and the cloud generates recommendations. A small cell cloud-enhanced eNode B, which functions as a fog device, analyzes indoor health status. The fog device stores the created recommendations, giving consumers timely and low-latency recommendations.

3.5.3 Smart homes and automation

A paradigm for creating voice-activated home automation systems that enhance comfort and independence is presented in this study. The paradigm emphasizes context-aware decision-making, which necessitates using actual sensor data to deduce unclear facts. Communication is reasoned using the same concepts and relations using a Markov logic network technique and a hierarchical knowledge model (Chahuara, Portet, and Vacher 2017). The framework leverages a priori knowledge to acquire weights and structure from data, addressing the problem of adapting expert systems to new applications.

Building a smart house isn't as easy as it sounds; it requires a variety of tools and technology. Therefore, the concept of context-aware low-power intelligent SmartHome (CLPiSmartHome) is introduced. The communication paradigm for CLPiSmartHome is a shared communication medium or the same communication language (Khan et al. 2016). In addition, an architecture that converts any electronic device into a single platform service for intercommunication is presented. This model is the source of the potential application, evaluation, and results of the CLPiSmartHome design.

The smart home is the central point for IoT applications. It is a networked house supplying goods and services through the internet. However, privacy and security issues are raised if the smart home is controlled remotely. The IoT's rapid and massive technological advancement creates new requirements for secure and safe services, privacy protection, and efficient management of the smart home. The literature (Almusaylim and Zaman 2019) dwells on the smart home strategies while also pointing out the open issues and obstacles to overcome them.

3.5.4 Industrial IoT

By shifting computationally demanding tasks from machine-type devices to edge servers, edge computing offers a possible solution for the Industrial-IoT (IIoT). However, context unawareness, capacity-constrained batteries, and limited spectrum resources may jeopardize performance enhancement. To optimize long-term throughput while taking energy budget and service dependability constraints into account, this paper suggests a learning-based channel selection approach. The framework achieves assured performance with a restricted variation from ideal performance by utilizing matching theory, Lyapunov optimization, and ML (Liao et al. 2019). Simulations are used to confirm its dependability and efficacy.

Industry 4.0, a developing idea utilizing IoT technology for quicker, more effective manufacturing and administration procedures, is covered in the article. With an emphasis on ambient intelligence and context awareness, Industry 4.0 is propelling the spread of smart goods and services (Bisio et al. 2018). To illustrate the usefulness of this new framework, the essay also provides three examples of I4.0 applications in smart homes, smart factories, and health scenarios.

An industrial IoT context-aware information system that supports supervisors and operators, whether stationary or mobile, in manufacturing shop floors is presented. The system incorporates important IIoT ideas such as context modeling, service-oriented design, and multi-layering (Alexopoulos et al. 2018). This work applies to a variety of manufacturing settings, but it is especially pertinent to dynamic, semi-structured settings where employees are mobile and not confined to fixed workstations. To collect industry requirements and use cases for the tailored implementation of the suggested solution, a case study involving a medium-sized shipyard was carried out.

Numerous cybersecurity risks affect the IIoT platform (Alexopoulos et al. 2018), which consists of edge computer systems, modems, and communication controllers. A study offers a thorough risk analysis of the sector, concentrating on robotics, big data, and analytics-based technologies (Tariq et al. 2020). Additionally, the study uses a CNN and hazard and operability study (HAZOP) (Tariq et al. 2020) to assess operability and risk. The suggested endpoint security plan enhances malware detection, denial of service, and data breaches. Firmware-driven emulation increases data protection, enhances incident detection, and lowers breach ratios.

3.5.5 Autonomous vehicles (AVS)

Although autonomous cars are becoming more common in places with sophisticated infrastructure, they still require more contextual awareness to function in congested regions and during emergencies. There is a dearth of literature on context-aware systems, and their realization requires a systematic approach. The research (Fernandez-Rojas et al. 2019) employs a human-AVS (HAVS) perspective to classify pertinent material into two groups: HAVSs and contextual awareness about vehicle infrastructure. Ad-hoc networks, services, wireless communication, radio systems, and cybersecurity issues are all included in the former, which focuses on the communication backbone for AVSs.

Because of their rich texture information, camera sensors are essential to autonomous driving and driver assistance systems. Several 3D object detection methods have recently been developed as a result of deep learning techniques. To enhance 3D object detection, the examination of existing methods suggests a novel feature aggregation technique (Zhou et al. 2022).

Although robust perception across many driving circumstances is still a difficulty, AVs are predicted to transform transportation. Existing techniques, such as early and late fusion, are inadequate because of how rigidly they are implemented. HydraFusion (Malawade, Mortlock, and Al Faruque 2022)is a selective sensor fusion framework that learns the driving environment and fuses the optimal set of sensors to enhance robustness without sacrificing efficiency and is suggested as a solution to these drawbacks.

The approach for building a runtime monitor with environmental context awareness for intelligent transport systems (ITS), a subclass of cyber-physical systems (CPS), is presented in this study. The technique (Zhang, Xu, et al. 2024)is used and validated on both simulated and real-world objects, demonstrating that the monitors produced can efficiently track dangerous occurrences in various settings, enhancing the security of intelligent driving systems. In dynamic surroundings, autonomous cars encounter safety-critical situations that call for strategies to deal with them. Hazardous scenarios can arise from random failures or malfunctions, and the severity of these events varies depending on the operational context. Understanding the operational context of the vehicle is essential to ensuring safe conduct. The preliminary work for context-awareness-assisted safety is presented in this paper Haupt and Liggesmeyer 2022). It includes perspective, modeling, goal, and relevance-based categorization, and the proposal of a context meta-model and its essential components for creating a context model that is relevant to safety.

3.6 CHALLENGES IN CONTEXT-AWARE ENABLING TECHNOLOGIES

In the IoT space, context-aware systems are essential for making choices in real-time and providing individualized experiences. However, issues with accuracy, scalability, interoperability, privacy and security, and energy consumption arise when they are implemented as shown in Figure 3.4.

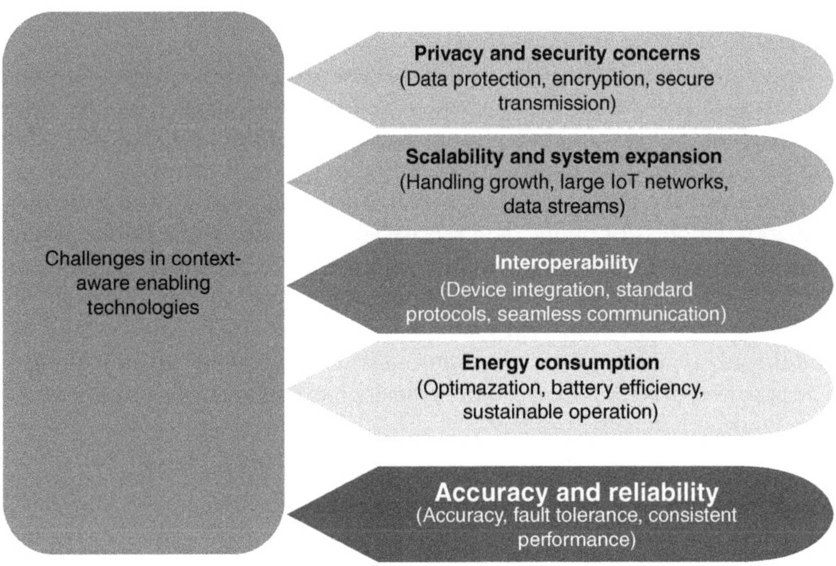

Figure 3.4 Challenges in context-aware enabling technologies.

3.6.1 Privacy and security concerns

To improve privacy via local information privacy (LIP) (Jiang et al. 2021), a context-aware privacy concept that utilizes prior knowledge to relax the standard local differential privacy (LDP). LIP, like other context-aware privacy concepts, gives more robust instance-wise privacy protection. Thus, it propagates the properties of a model of uncertainty in the setting of incomplete prior knowledge, post-processing, linkage, composability, and transferability. A broad approach to balancing utility and privacy for LIP-based privacy-protecting methods (Jiang et al. 2021) for both discrete and continuous-valued data types is also given. Examining three classes of perturbation processes, namely, randomized response (RR), random sampling (RS), and additive noise (AN), privacy techniques based on perturbation improve utility by incorporating previous knowledge into the perturbation parameters and to explore the utility-privacy tradeoffs offered by different methods and to provide the benefit of context-awareness.

Work based on information theory puts forth a mathematical scheme for coping with privacy in network caching systems in this paper. A server, a cache-assisted end user, and a foolproof content caching method that can maximize privacy protection while still keeping track of the average delivery load are elements of the model of the paradigm. Moreover, the suggested ε-constraint optimization approach finds the highest degree of privacy that can be achieved for any given average delivery load. Context-oriented privacy model (Hassanpour et al. 2020) is proven right by numerical results and comparisons, and also indicates that better caching methods are needed to prevent adversaries from getting hold of the most important contextual data.

Sequence information privacy (SIP) is a new privacy concept that considers the intrinsic correlation of data and also guarantees privacy for the entire data stream. SIP is a lightweight, modular mechanism designed and offers privacy protection comparable to LDP (Jiang, Li, and Tandon 2024). It also proposes privacy-preserving data perturbation methods, as well as investigating two models of online data release. To demonstrate the utility-privacy tradeoff offered by SIP that is in line with other research, the authors have carried out experiments with real datasets. The results clearly show that SIP is still superior to LDP-based methods in terms of utility-privacy tradeoff, and the advantage even becomes more distinct for smaller privacy budgets. The paper reports that SIP is a more efficient method of publishing streaming data while still keeping privacy.

3.6.2 Scalability and system expansion

Real-time health monitoring systems that employ the IoT have drawn interest in lowering the death rate from heart failure, a leading cause of death worldwide. These systems' reaction time, scalability, latency, and fault tolerance are some of their drawbacks. To construct healthcare systems, a four-layer hierarchical architecture is suggested (Borujeni, Fathy, and Mozayani 2019). The model sends patient vital signs to a smart healthcare system via a network of body sensors. Because each layer corresponds to a distinct level of heart failure, uncomplicated and low-risk heart failure can be promptly identified before it progresses to severe levels.

Preventing noncommunicable diseases requires a healthy diet and regular exercise, but both might be difficult to achieve because of social and environmental factors. Even though there are many tools and apps available to assist patients, they are insufficient on their own. To assist patients in overcoming health challenges, strategies must integrate data, people, and technology. To improve healthcare, a scalable strategy that combines technology, data, and people was proposed (Vianna and Barbosa 2019). The authors propose a recommendation technique that is based on clustering to scale the collaborative filtering (CF) algorithms (Datta et al. 2015) beyond limitations. The main goal is to cut down on running time but still keep the suggestion quality intact, thus ensuring scalability and capability of larger datasets. The algorithm has proved to be very effective in cutting down on running time by being tested on the MovieLens dataset.

3.6.3 Interoperability

The approach to interoperability in context-aware software systems highlights the particularly good aspects of it as well as the less good ones. After the differences that may occur due to the development platforms, data formats, culture, and legal problems are solved, the framework (Motta, de Oliveira, and Travassos 2019) draws a picture of how things could work together for a certain goal. Interchangeable behavioral principles, such as the assessment techniques, the problems, the concerns, and the advantages, may be used to measure, improve, and monitor the framework. The framework can improve interoperability and also lead the way in the development of software systems.

It was proposed that the Ontology as a Service (OaaS) (Barangi et al. 2023). A mechanism should be employed to improve semantic interoperability in context-aware systems. OaaS approach through a stand-alone ontology service alleviates the problem of semantic interoperability. The service encompasses repositories, a web service that represents the correspondences among existing context ideas, and a web service that introduces context concepts into already existing ontologies. A case study was presented to test the approach's performance in solving semantic conflicts between two IoT applications taken from smart homes and the health care sector.

Emphasizing SandS, a social network ecosystem, the discussion about the user modeling, context awareness, and content adaptation in Smart Home environments. The research that highlights the implementation of fuzzy stereotypical profiles (Siolas et al. 2013) to represent users and their relationships, the employment of knowledge-based expert mappings to achieve efficient interoperability between users and system services, even among languages with different semantic expressiveness, and context modeling in pervasive computing systems. The work aims to detail the innovative user experience in Smart Home environments from a semantics perspective.

3.6.4 Energy consumption

The non-causal adaptive context-aware bidirectional deep learning model for energy disaggregation, bidirectional long short-term memory (BiLSTM), is presented in this research (Kaselimi et al. 2020). Deep recurrent long short-term memory (DR-LSTM) neural networks are used in the model to take seasonality and other contextual factors into account, as well as to adjust for non-causality. The optimal configuration is chosen using a Bayesian-optimized framework that is powered by an adaptive self-training process.

Hardware's energy consumption is considerably affected by software systems. It is especially true in cyber-physical systems, where battery life is strongly influenced by power usage. The software developers should be aware of the energy consumed, and they need to create systems that change their behavior to reduce energy consumption. This can be achieved, for instance, by monitoring system utilization scenarios and providing runtime assistance to respond to changes that impact the energy footprint. The model that deals with a cyber-physical system focuses on software adaptation and demonstrates how to generate context-aware, energy-efficient applications through the use of modern software engineering techniques such as separation of concerns and dynamic software product lines (Horcas Aguilera, Pinto, and Fuentes 2019).

An ML-based algorithm that manages resource use (Nawrocki and Sniezynski 2020) learns the context and uses ML algorithms to lower the energy consumption of a mobile device. The most important thing for mobile systems is the fact that the system learns how to distribute services and tasks between the device and the cloud; thus, it can distribute resources correctly, and this will be the most efficient. While taking a decision, contextual elements such as the type of network needed, location, possible duration, and cost of running the application or service are considered. For this, a service selection algorithm and supervised learning agent architecture are suggested. Software development and testing were the benchmarks the system was evaluated, showing that more tasks can be assigned efficiently to cloud resources or mobile devices.

3.6.5 Accuracy and reliability

Sensors are used by context-aware systems to gather contextual data from their surroundings, which informs control choices. On the other hand, opponents can control the environment to provide them with advantages they wouldn't otherwise have. By adding sentience-like components to the proximity-based access control model, the expansion of context-aware system to secure enterprise content (CASSEC) to solve this issue was carried out (Zhu et al. 2017). This makes it possible to conclude the credibility of the information that has been gathered. By introducing two new authentication methods—co-proximity authentication using Bluetooth Low Energy beacons and biometric authentication with accelerometer and fingerprint sensors—CASSEC 2.0 assesses confidence structures.

Two issues that plague traditional crowdsourcing systems are breaking down large jobs into smaller, more manageable tasks that necessitate the capacity of requesters to decompose, which may cause them to quit the project. Furthermore, assignments are sometimes given to several employees in duplicate, which might result in unreliability when there are a lot of dishonest employees. The exploration of crowdsourcing (Jiang et al. 2017) in social networks introduces a unique, trustworthy, context-aware crowdsourcing technique. More requesters can complete complicated tasks because of the method's autonomous coordination between contextual workers in the social network and allocated workers. The reputations of contextual workers in the social network serve as a reliable indicator of a worker's dependability, therefore mitigating the unreliability of temporary or malevolent workers. The method outperforms earlier benchmark techniques in terms of job allocation and execution efficiency, and it works with both simplex and multiplex networks.

To overcome the difficulties presented by black-box services, the research suggests common address redundancy protocol (CARP), a novel context-aware reliability prediction technique (Oluwatimi, Damiani, and Bertino 2018) that gives online reliability prediction results by utilizing user usage history to build context-aware reliability models. The data sparsity issue that restricts the prediction accuracy of other current methods is addressed by CARP.

3.7 FUTURE DIRECTIONS

The union of Blockchain and IoT presents life-changing opportunities for enhancing solitude, security, and adeptness, but the potential allure can only be fully realized through persistent tests. A critical course includes toughening privacy designs to meet the increasing complexity of IoT ecosystems. Advanced cryptographic methods, in the way as ZKP and homomorphic encryption, will be necessary for permissive secure transactions and dossier giving without revealing sensitive facts. The most essential approach is focusing on reducing disclosures and off-chain repositories that minimize the limitations of the Blockchain. The evolution of hybrid methodologies of integrating private and public Blockchain frameworks creates a bridge for balancing transparency preservation and maintenance of confidentiality in data. The interoperability factor that establishes the Layer-2 security between the Blockchain framework and IoT leads to the creation of unified methodologies.

The integration of AI for the analysis of prediction and finding inconsistent behavior enables the Blockchain to focus mainly on major obstacles and helps in developing IoT functionalities. Another important factor is efficiency, which is getting affected by restricted computational capacity and cost matters. The algorithms, such as Delegated Proof of Stake (D-PoS) and PBFT, are known as lightweight harmony algorithms that are responsible for the security and efficiency of the system.

Ethical and permissible considerations are evenly fault-finding in forming the future of Blockchain-IoT integration. Adherence to all-encompassing dossier protection management, in a way that General Data Protection Regulation (GDPR), Central Consumer Protection Authority (CCPA), and Health Insurance Portability and Accountability Act (HIPAA) must be a preference. Incorporating privacy-by-design laws into Blockchain foundations will guarantee compliance while supporting trust with users. Ethical dossier management practices, containing transparency in dossier habits and informed consumer consent, are essential for safeguarding consumer rights and preventing misuse. Governance foundations must likewise address accountability in dispersed schemes, building mechanisms for dispute determination and advancing mature behavior between colleagues.

3.8 CONCLUSION

The integration of Blockchain science accompanying IoT shows a paradigm shift in how dossier privacy, safety, and scalability are governed in progressively interconnected environments. Blockchain's scattered nature, linked to accompanying allure features like transparency, pseudonymity, and continuity, provides a strong foundation for sending traditional IoT exposures, in the way that dossier breaches, unauthorized approaches, and lack of trust between stakeholders. The capability to follow the path and substantiate device interplays outside relying on concentrated experts not only supports a more secure environment but also reinforces operational effectiveness across activities. The differing Blockchain architectures—public, private, and consortium—cater to diverse use cases, and permissive, tailor-made resolutions for IoT applications in healthcare, supply chain administration, smart downtowns, and beyond. Smart contracts further improve this environment by automating processes, guaranteeing data integrity, and lowering operational overhead.

This physiognomy explains the huge potential of Blockchain in redefining IoT applications by making the ruling class more secure, understandable, and user-friendly. Blockchain's inherent disadvantages, such as scalability issues, extreme energy consumption, and abeyance, need to be talked about to ensure unity accompanying the resource-intensive character of IoT instruments. Innovations in consensus algorithms, inconsequential cryptographic answers, and hybrid Blockchain foundations are essential to defeating these barriers. Furthermore, as the IoT environment expands, the abrupt capacity of data creation will necessitate the development of state-of-the-art off-chain storage solutions and data condensation methods to maintain method adeptness and approachability.

Looking ahead, the righteous and allowable implications of Blockchain maintenance must wait for a focus. Ensuring compliance with accompanying dossier protection rules to a degree, GDPR, HIPAA, and CCPA are owned by fostering trust with consumers and shareholders. Governance frameworks that delimit accountability, maturity, and dispute-determination devices will be critical in directing distributed systems. Ethical concerns, in the way as cognizant user consent, understandable dossier usage, and impartial approach, must again guide the development and arrangement of Blockchain-based IoT structures to avoid misuse and guarantee justice. The future of Blockchain in IoT is also firmly relies on progress in complementary sciences. The unification of AI accompanying Blockchain can allow predictive analytics, danger discovery, and optimized administration, making IoT orders smarter and more adaptive. Similarly, the arrival of quantity estimating presents both moments and challenges. While it has the potential to transform Blockchain's computational capabilities, it requires the use of more existing cryptographic standards, making it necessary for a fully holistic development of quantum-resistant encryption arrangements.

Sensing, modeling, AI/ML, communication protocols, and computing paradigms are the enabling technologies for context-aware systems that combine to produce intelligent, adaptive IoT applications. The basis is provided by sensing and data collection, while smooth analysis and interaction are made possible by modeling, AI, and communication protocols. Future developments will resolve security issues, increase energy efficiency, and boost scalability. Real-time, context-aware solutions will become possible when AI advancements are combined with IoT capabilities. These developments will lead to broader applications in various fields, such as industrial automation, smart cities, and healthcare.

However, the existence of limitations, the Blockchain-based context-aware technologies stand out for functionality enhancement and improved customization in IoT-based systems. For safeguarding data privacy, security mechanisms and privacy-preserving methodologies are carried out. For maintaining the scalability and performance of the system on a long-term basis, the consumption of energy by the system needs to be monitored regularly. These conditional factors help the system to be secure, efficient, versatile, more scalable, and dependable for securing private data.

REFERENCES

Ahmed, Ahmed Abdelmoamen, and Oluwayemisi Alabi. 2024. "Secure and scalable blockchain-based federated learning for cryptocurrency fraud detection: A systematic review." *IEEE Access* 12:102219–102241.

Akpakwu, Godfrey A., Gerhard P. Hancke, and Adnan M. Abu-Mahfouz. 2020. "CACC: Context-aware congestion control approach for lightweight CoAP/UDP-based Internet of Things traffic." *Transactions on Emerging Telecommunications Technologies* 31 (2):e3822.

Al-Hammuri, Khalid, Fayez Gebali, and Awos Kanan. 2024. "ZTCloudGuard: Zero trust context-aware access management framework to avoid medical errors in the era of generative AI and cloud-based health information ecosystems." *AI* 5 (3):1111–1131.

Alexopoulos, Kosmas, Konstantinos Sipsas, Evangelos Xanthakis, Sotiris Makris, and Dimitris Mourtzis. 2018. "An industrial internet of things based platform for context-aware information services in manufacturing." *International Journal of Computer Integrated Manufacturing* 31 (11):1111–1123.

Almusaylim, Zahrah A, and Noor Zaman. 2019. "A review on smart home present state and challenges: linked to context-awareness internet of things (IoT)." *Wireless Networks* 25:3193–3204.

Barangi, Hamed, Shekoufeh Kolahdouz Rahimi, Bahman Zamani, and Hossein Moradi. 2023. "An ontology-based approach to facilitate semantic interoperability of context-aware systems." 2023 28th International Computer Conference, Computer Society of Iran (CSICC).

Bisio, Igor, Chiara Garibotto, Aldo Grattarola, Fabio Lavagetto, and Andrea Sciarrone. 2018. "Exploiting context-aware capabilities over the internet of things for industry 4.0 applications." *IEEE Network* 32 (3):101–107.

Bormann, Carsten, Angelo P Castellani, and Zach Shelby. 2012. "Coap: An application protocol for billions of tiny internet nodes." *IEEE Internet Computing* 16 (2):62–67.

Borujeni, Ahmad Malekian, Mahmood Fathy, and Nasser Mozayani. 2019. "A hierarchical, scalable architecture for a real-time monitoring system for an electrocardiography, using context-aware computing." *Journal of Biomedical Informatics*, 96: 103251–103264.

Chahuara, Pedro, François Portet, and Michel Vacher. 2017. "Context-aware decision making under uncertainty for voice-based control of smart home." *Expert Systems with Applications* 75:63–79.

Datta, Suman, Joydeep Das, Prosenjit Gupta, and Subhashis Majumder. 2015. "SCARS: A scalable context-aware recommendation system." Proceedings of the 2015 Third International Conference on Computer, Communication, Control and Information Technology (C3IT).

Daye, Dania, Regina Parker, Satvik Tripathi, Meredith Cox, Sebastian Brito Orama, Leonardo Valentin, Christopher P. Bridge, and Raul N. Uppot. 2024. "CASCADE: Context-aware data-driven AI for streamlined multidisciplinary tumor board recommendations in oncology." *Cancers* 16 (11):1975.

Elayan, Haya, Moayad Aloqaily, and Mohsen Guizani. 2021. "Digital twin for intelligent context-aware IoT healthcare systems." *IEEE Internet of Things Journal* 8 (23):16749–16757.

Fernandez-Rojas, Raul, Anthony Perry, Hemant Singh, Benjamin Campbell, Saber Elsayed, Robert Hunjet, and Hussein A Abbass. 2019. "Contextual awareness in human-advanced-vehicle systems: A survey." *IEEE Access* 7:33304–33328.

Ghosh, Shreya, Jaydeep Das, Soumya K Ghosh, and Rajkumar Buyya. 2020. "CLAWER: Context-aware cloud-fog based workflow management framework for health emergency services." 2020 20th IEEE/ACM International Symposium on Cluster, Cloud and Internet Computing (CCGRID).

Goyat, Rekha, Gulshan Kumar, Mamoun Alazab, Mauro Conti, Mritunjay Kumar Rai, Reji Thomas, Rahul Saha, and Tai-Hoon Kim. 2020. "Blockchain-based data storage with privacy and authentication in internet of things." *IEEE Internet of Things Journal* 9 (16):14203–14215.

Goyat, Rekha, Gulshan Kumar, Mauro Conti, Tannishtha Devgun, Rahul Saha, and Reji Thomas. 2023. "BENIGREEN: Blockchain-based energy-efficient privacy-preserving scheme for green IoT." *IEEE Internet of Things Journal* 10 (18):16480–16493.

Hassan, Mahmood Ul, Amin A Al-Awady, Abid Ali, Muhammad Munwar Iqbal, Muhammad Akram, and Harun Jamil. 2024. "Smart resource allocation in mobile cloud next-generation network (NGN) orchestration with context-aware data and machine learning for the cost optimization of microservice applications." *Sensors* 24 (3):865.

Hassanpour, Seyedeh Bahereh, Abolfazl Diyanat, Ahmad Khonsari, Seyed Pooya Shariatpanahi, and Aresh Dadlani. 2020. "Context-aware privacy preservation in network caching: An information theoretic approach." *IEEE Communications Letters* 25 (1):54–58.

Haupt, Nikita Bhardwaj, and Peter Liggesmeyer. 2022. "Towards context-awareness for enhanced safety of autonomous vehicles." Intelligent Systems and Applications: Proceedings of the 2021 Intelligent Systems Conference (IntelliSys) Volume 2.

Horcas Aguilera, José Miguel, Mónica Pinto, and M.Y. Fuentes. 2019. "Context-aware energy-efficient applications for cyber-physical systems." *Ad Hoc Networks* 82 (January 2019):15–30.

Hu, Zhongxu, Yuxin Cai, Qinghua Li, Kui Su, and Chen Lv. 2024. "Context-aware driver attention estimation using multi-hierarchy saliency fusion with gaze tracking." *IEEE Transactions on Intelligent Transportation System* 25 (8):8602–8614.

Huang, Haosheng, Yi Cheng, Weihua Dong, Georg Gartner, Jukka M Krisp, and Liqiu Meng. 2024. "Context modeling and processing in location based services: Research challenges and opportunities." *Journal of Location Based Services 18:*1–27.

Hussain, Muzammil, Waheed Javed, Owais Hakeem, Abdullah Yousafzai, Alisha Younas, Mazhar Javed Awan, Haitham Nobanee, and Azlan Mohd Zain.

2021. "Blockchain-based IoT devices in supply chain management: A systematic literature review." *Sustainability* 13 (24):13646.

Islam, Mir Salim Ul, Ashok Kumar, and Yu-Chen Hu. 2021. "Context-aware scheduling in Fog computing: A survey, taxonomy, challenges and future directions." Journal of Network and Computer Applications 180:103008.

Jenevain, Matheus B, Milena F Pinto, Mario AR Dantas, Regina MMB Villela, Jose MN David, and Victor SA Menezes. 2024. "Enabling intelligent data exchange in the Brazilian energy sector: A context-aware ontological approach." International Conference on Advanced Information Networking and Applications.

Jiang, Bo, Ming Li, and Ravi Tandon. 2024. "Online context-aware streaming data release with sequence information privacy." *IEEE Transactions on Information Forensics and Security* 19:4390–4405.

Jiang, Bo, Mohamed Seif, Ravi Tandon, and Ming Li. 2021. "Context-aware local information privacy." *IEEE Transactions on Information Forensics and Security* 16:3694–3708.

Jiang, Jiuchuan, Bo An, Yichuan Jiang, and Donghui Lin. 2017. "Context-aware reliable crowdsourcing in social networks." *IEEE Transactions on Systems, Man, and Cybernetics: Systems* 50 (2):617–632.

Kaselimi, Maria, Nikolaos Doulamis, Athanasios Voulodimos, Eftychios Protopapadakis, and Anastasios Doulamis. 2020. "Context aware energy disaggregation using adaptive bidirectional LSTM models." *IEEE Transactions on Smart Grid* 11 (4):3054–3067.

Kaur, Amanpreet, Gurpreet Singh, Vinay Kukreja, Sparsh Sharma, Saurabh Singh, and Byungun Yoon. 2022. "Adaptation of IoT with blockchain in food supply chain management: An analysis-based review in development, benefits and potential applications." *Sensors* 22 (21):8174.

Khan, Murad, Sadia Din, Sohail Jabbar, Moneeb Gohar, Hemant Ghayvat, and SC Mukhopadhyay. 2016. "Context-aware low power intelligent smart home based on the Internet of things." *Computers & Electrical Engineering* 52:208–222.

Kim, Geonwoo, Seongju Kang, Jiwoo Park, and Kwangsue Chung. 2019. "An MQTT-based context-aware autonomous system in oneM2M architecture." *IEEE Internet of Things Journal* 6 (5):8519–8528.

Kolbinger, Fiona R., Sebastian Bodenstedt, Matthias Carstens, Stefan Leger, Stefanie Krell, Franziska M Rinner, Thomas P Nielen, Johanna Kirchberg, Johannes Fritzmann, and Jürgen Weitz. 2024. "Artificial Intelligence for context-aware surgical guidance in complex robot-assisted oncological procedures: An exploratory feasibility study." *European Journal of Surgical Oncology* 50 (12):106996.

Li, Shengdong, Chen Gong, Yuqing Zhu, Chuanwen Luo, Yi Hong, and Xueqiang Lv. 2024. "Context-aware multi-level question embedding fusion for visual question answering." *Information Fusion* 102:102000.

Li, Xiaomeng, Fei Li, Yanyu Qi, and Zhenbo Li. 2024. "Plantformer: Plant point cloud completion based on local–global feature aggregation and spatial context-aware transformer." *Neural Computing and Applications* 37:2747–2762.

Li, Yangyang, Yuelin Li, Shihuai Zhang, Guangyuan Liu, Yanqiao Chen, Ronghua Shang, and Licheng Jiao. 2024. "An attention-based, context-aware multimodal fusion method for sarcasm detection using inter-modality inconsistency." *Knowledge-Based Systems* 287(c): 111457–111458.

Liao, Haijun, Zhenyu Zhou, Xiongwen Zhao, Lei Zhang, Shahid Mumtaz, Alireza Jolfaei, Syed Hassan Ahmed, and Ali Kashif Bashir. 2019. "Learning-based context-aware resource allocation for edge-computing-empowered industrial IoT." *IEEE Internet of Things Journal* 7 (5):4260–4277.

Liu, Yu, Tong Li, Runzi Zhang, Zhao Jin, Mingkai Tong, Wenmao Liu, Yiting Wang, and Zhen Yang. 2024. "A context-aware clustering approach for assisting operators in classifying security alerts." *IEEE Transactions on Software Engineering* 51:153–171.

Malawade, Arnav Vaibhav, Trier Mortlock, and Mohammad Abdullah Al Faruque. 2022. "HydraFusion: Context-aware selective sensor fusion for robust and efficient autonomous vehicle perception." 2022 ACM/IEEE 13th International Conference on Cyber-Physical Systems (ICCPS).

Motta, Rebeca C, Káthia M de Oliveira, and Guilherme H Travassos. 2019. "A conceptual perspective on interoperability in context-aware software systems." *Information and Software Technology* 114:231 257.

Nawrocki, Piotr, and Bartlomiej Sniezynski. 2020. "Adaptive context-aware energy optimization for services on mobile devices with use of machine learning." *Wireless Personal Communications* 115 (3):1839–1867.

Oluwatimi, Oyindamola, Maria Luisa Damiani, and Elisa Bertino. 2018. "A context-aware system to secure enterprise content: Incorporating reliability specifiers." *Computers & Security* 77:162–178.

Padma, Adla, and Mangayarkarasi Ramaiah. 2024. "Blockchain based an efficient and secure privacy preserved framework for smart cities." *IEEE Access* 12:21985–22002.

Qi, Lianyong, Chunhua Hu, Xuyun Zhang, Mohammad R Khosravi, Suraj Sharma, Shaoning Pang, and Tian Wang. 2020. "Privacy-aware data fusion and prediction with spatial-temporal context for smart city industrial environment." *IEEE Transactions on Industrial Informatics* 17 (6):4159–4167.

Rahmadika, Sandi, Philip Virgil Astillo, Gaurav Choudhary, Daniel Gerbi Duguma, Vishal Sharma, and Ilsun You. 2022. "Blockchain-based privacy preservation scheme for misbehavior detection in lightweight IoMT devices." *IEEE Journal of Biomedical and Health Informatics* 27 (2):710–721.

Reddy, K Hemant Kumar, Ashish Kr Luhach, Buddhadeb Pradhan, Jatindra Kumar Dash, and Diptendu Sinha Roy. 2020. "A genetic algorithm for energy efficient fog layer resource management in context-aware smart cities." *Sustainable Cities and Society* 63:102428.

Sai, Siva, Vikas Hassija, Vinay Chamola, and Mohsen Guizani. 2023. "Federated learning and NFT-based privacy-preserving medical data sharing scheme for intelligent diagnosis in smart healthcare." IEEE Internet of Things Journal 11(4):5568–5577.

Schürholz, Daniel, Sylvain Kubler, and Arkady Zaslavsky. 2020. "Artificial intelligence-enabled context-aware air quality prediction for smart cities." *Journal of Cleaner Production* 271:121941.

Shen, Xiaodong, Chang Xu, Liehuang Zhu, Rongxing Lu, Yunguo Guan, and Xiaoming Zhang. 2023. "Blockchain-based lightweight and privacy-preserving quality assurance framework in crowdsensing systems." *IEEE Internet of Things Journal* 11 (1):974–986.

Singh, Rajani, Ashutosh Dhar Dwivedi, Gautam Srivastava, Pushpita Chatterjee, and Jerry Chun-Wei Lin. 2023. "A privacy-preserving Internet of Things smart healthcare financial system." IEEE Internet of Things Journal 10 (21):18452–18460.

Siolas, Giorgos, George Caridakis, Phivos Mylonas, Stefanos Kollias, and Andreas Stafylopatis. 2013. *"Context-aware user modeling and semantic interoperability in smart home environments."* 2013 8th International Workshop on Semantic and Social Media Adaptation and Personalization.

Tariq, Usman, Ahmad O Aseeri, Mohammed Saeed Alkatheiri, and Yu Zhuang. 2020. "Context-aware autonomous security assertion for industrial IoT." *IEEE Access* 8:191785–191794.

Temenos, Anastasios, Nikos Temenos, Ioannis N Tzortzis, Ioannis Rallis, Anastasios Doulamis, and Nikolaos Doulamis. 2024. "C2A-DC: A context-aware adaptive data cube framework for environmental monitoring and climate change crisis management." *Remote Sensing Applications: Society and Environment* 34:101171.

Vianna, Henrique Damasceno, and Jorge Luis Victória Barbosa. 2019. "A scalable model for building context-aware applications for noncommunicable diseases prevention." *Information Processing Letters* 148:1–6.

Wang, Jiacheng, Hongyang Du, Zengshan Tian, Dusit Niyato, Jiawen Kang, and Xuemin Shen. 2023. "Semantic-aware sensing information transmission for metaverse: A contest theoretic approach." *IEEE Transactions on Wireless Communications* 22 (8):5214–5228.

Zhang, Jianming, Yufan He, Wentao Chen, Li-Dan Kuang, and Bin Zheng. 2024. "CorrFormer: Context-aware tracking with cross-correlation and transformer." *Computers and Electrical Engineering* 114:109075.

Zhang, Yu, Sijie Xu, Hongyi Chen, Uzair Aslam Bhatt, and Mengxing Huang. 2024. "Context-aware environment online monitoring for safety autonomous vehicle systems: An automata-theoretic approach." *Journal of Cloud Computing* 13 (1):6.

Zhang, Yuchao, Xiaotian Wang, Xiaofeng He, Ning Zhang, Zibin Zheng, and Ke Xu. 2023. "Galaxy: A scalable BFT and privacy-preserving pub/sub IoT data sharing framework based on blockchain." *IEEE Internet of Things Journal* 11:5222–5236.

Zhou, Dingfu, Xibin Song, Jin Fang, Yuchao Dai, Hongdong Li, and Liangjun Zhang. 2022. "Context-aware 3D object detection from a single image in autonomous driving." *IEEE Transactions on Intelligent Transportation Systems* 23 (10):18568–18580.

Zhu, Jieming, Pinjia He, Qi Xie, Zibin Zheng, and Michael R Lyu. 2017. "CARP: context-aware reliability prediction of black-box web services." *2017 IEEE International Conference on Web Services (ICWS), pp. 1–17.*

Chapter 4

Developing a generic framework for Blockchain-based applications for IoT

Design, implementation, and best practices

Smita Dange, Suhas Lawand, and Prashant Nitnaware

4.1 INTRODUCTION

4.1.1 Overview of IoT and Blockchain integration

The rapid evolution of the Internet of Things (IoT) has resulted in a surge of interconnected devices, generating enormous volumes of data and enabling automation across various sectors, such as healthcare, smart cities, industrial automation, and supply chain management. However, IoT systems face significant challenges due to their inherent characteristics, including decentralized data collection, limited device resources, and various security vulnerabilities. These challenges are often not effectively addressed by traditional centralized solutions, which struggle to provide scalability, security, and efficient management of large-scale distributed networks.

Blockchain technology, as highlighted by Dange and Chatterjee (2020), offers a robust solution to many of these challenges by providing a decentralized, immutable, and transparent framework for managing IoT data. By integrating Blockchain with IoT, several critical issues can be effectively addressed, such as data integrity, secure device authentication, and privacy. The decentralized Blockchain ledger ensures that the data transmitted between devices remains tamper-proof, providing a high level of trust for all transactions in an IoT network. This immutability is crucial in industries such as healthcare, where data integrity and security are paramount for sensitive information like medical records and patient health data.

Furthermore, Blockchain improves IoT security through consensus mechanisms that validate transactions across distributed nodes, reducing the risk of central points of failure and improving the resilience of the entire system. The integration of smart contracts further automates actions based on predefined rules, allowing seamless operations without the need for intermediaries. This is particularly beneficial in the supply chain management and medical services sectors, where automation and efficiency are

DOI: 10.1201/9781003584728-4

essential to reduce costs and improve operational results. Smart contracts allow secure and transparent execution of transactions and processes, such as verifying the authenticity of goods in the supply chain or automating medical procedures in healthcare systems.

Despite the promising potential, the integration of Blockchain with IoT is not without challenges. Issues such as scalability, energy efficiency, and latency must be addressed, especially considering the resource-constrained nature of many IoT devices. Recent advances in lightweight consensus algorithms, off-chain storage, and energy-efficient protocols are essential to overcome these challenges, enabling the seamless deployment of Blockchain solutions in IoT ecosystems. In addition, IoT systems require effective real-time data processing and secure communication channels to operate optimally, both of which can be enhanced through Blockchain integration.

A recent review by Fernandez-Carames and Fraga-Lamas (2024) confirms that the combination of IoT and Blockchain technologies holds the potential to revolutionize a wide range of industries by improving data management, ensuring privacy, and providing secure, automated solutions for device authentication and transaction validation. As research into Blockchain's scalability and optimization in IoT environments continues, these technologies will increasingly be adopted for a range of applications, from enhancing the security and privacy of healthcare IoT systems to enabling transparent and traceable supply chain operations. However, the full realization of this potential will require continued innovation in consensus mechanisms and energy-efficient protocols, making the intersection of IoT and Blockchain a dynamic and critical area for ongoing research and development.

4.1.2 Motivation for developing generic framework

The convergence of Blockchain and IoT has opened new avenues for improving security, decentralization, and data transparency in a wide range of domains, including healthcare, smart cities, and industrial systems. Despite these opportunities, the development of Blockchain-based IoT applications remains fragmented and challenging due to heterogeneity in devices, platforms, and protocols.

Currently, developers lack a unified framework to guide the integration process, often resulting in ad-hoc implementations with limited scalability, inefficient consensus mechanisms, and inconsistent security practices. Moreover, existing development efforts struggle to meet the real-time data handling and energy constraints imposed by IoT environments.

These gaps underscore the pressing need for a generic, adaptable, and structured development framework that can standardize design principles, improve interoperability, and streamline the deployment of Blockchain-enabled IoT systems. By establishing such a framework, this work aims to bridge the current disconnect between Blockchain capabilities and the demands of IoT applications, enabling more robust, efficient, and secure next-generation systems.

4.1.3 Contributions of the chapter

This chapter offers several key contributions toward the development of a robust framework for integrating Blockchain technology with IoT:

1. *Establishment of the need for a structured framework*: The chapter highlights the critical importance of a systematic approach to Blockchain IoT development, emphasizing the necessity of maintaining security, interoperability, and scalability across diverse IoT environments.
2. *Comprehensive analysis of integration challenges*: It provides an in-depth discussion of prevalent challenges, such as latency, limited computing resources, energy consumption, and security risks. The chapter outlines how the proposed framework addresses these limitations using best practices and modular design principles.
3. *Proposal of a generic and adaptable architecture*: A novel, generic framework is introduced that can be customized to various application domains. This flexibility allows developers to tailor Blockchain integration strategies while adhering to standard protocols and ensuring system resilience.
4. *Practical implementation guidelines*: Through the inclusion of detailed case studies such as smart healthcare and precision agriculture, the chapter demonstrates the real-world applicability of the proposed framework, offering actionable insights for researchers and industry practitioners.

By detailing these contributions, this chapter aims to serve as both a foundational reference for further academic inquiry and a practical guide for developers seeking to harness Blockchain for next-generation IoT systems.

4.1.4 Structure of the chapter

The chapter is structured to provide a comprehensive guide on integrating Blockchain with IoT systems. It begins with an introduction outlining the growing importance of this integration and the specific objectives of developing a flexible framework. Next, it covers fundamental concepts, exploring the architectures of IoT and Blockchain, followed by a discussion on the synergy between them. The chapter then delves into the requirements and design of a scalable and secure Blockchain-based IoT framework. Implementation strategies are detailed, including platform selection, smart contract development, and testing methodologies. Real-world case studies illustrate the practical application of the framework in diverse domains like supply chain management and healthcare. The chapter also addresses various challenges such as scalability, security, and regulatory issues, proposing viable solutions. It concludes with future trends, highlighting emerging technologies, evolving Blockchain protocols, and potential research directions, ending with a call to action for further exploration and adoption of the proposed framework.

4.2 FUNDAMENTAL CONCEPTS

4.2.1 IoT architecture

IoT envisions any time, anywhere, anyway, and anything. IoT should be viewed as the fundamental component of the current internet, and improvements should be made to realize this vision. Therefore, the idea of architecture was born. Architecture is a foundation technology that allows humans to layer on top of different items to connect and interact with them. The foundation of the IoT architecture is machine-to-machine (M2M) architecture.

IoT architecture refers to the overall structure and components of an IoT system. IoT system is a software and hardware combination to achieve the desired result. Processes, stores them, and retrieves and analyzes them. Communication systems play an important role in facilitating communication among various IoT components. Software, hardware, and communication systems act as the building blocks of an IoT architecture. As per Ray (2018), an effective IoT architecture is needed to achieve reliable and secure convergence of operational and information technology.

Depending on the requirements and tasks to be handled, the layers in the IoT architecture can vary. Various researchers have proposed three-layered, four-layered, five-layered, six-layered, and seven-layered architectures for the IoT. N. M. Kumar and Mallick (2018) mentioned that layers have been added to separate functionality and balance load. Typically, the architectural layer increases, as does the complexity of the design of the IoT system and its integration with the real-time environment. More interaction leads to more security issues and the mishandling of data mentioned by Lin et al. (2017) and Khan et al. (2012).

In light of this, the researcher considered the basic three-layered IoT architecture as shown in Figure 4.1. Basic IoT architecture is the most practical and is simple to implement. The architecture has three layers: the thing layer (perception layer), the edge layer (network layer), and the application layer. The role and functionality of each layer are as follows.

Thing layer: The perception layer is the lowest in IoT architecture, and its primary function is to collect data from the physical environment using

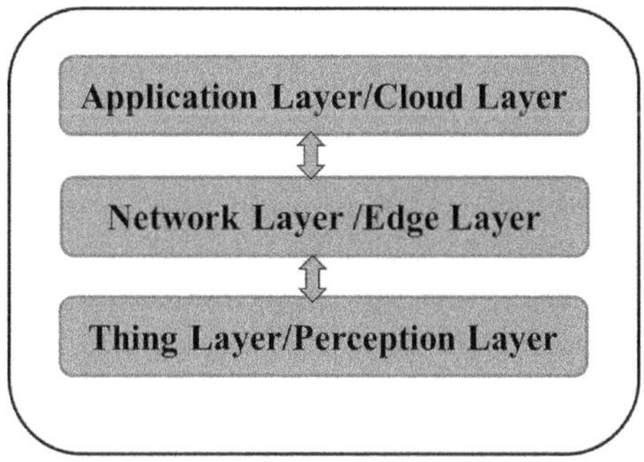

Figure 4.1 The three-layered IoT architecture.

sensors, devices, and actuators. This layer plays a crucial role in capturing and digitizing various types of data, such as temperature, pressure, humidity, motion, and location.

The perception layer consists of a wide range of devices that are responsible for collecting data. These devices include sensors, cameras, microphones, and radio frequency identification (RFID) tags. Each device is designed to capture specific types of data based on the application requirements. It is important to note that the processing capability of the perception layer is usually limited compared to the higher layers of the IoT architecture. The main focus of this layer is to gather data from the physical world and transmit it to the next layer for further processing and analysis.

The collected data from the perception layer serves as the foundation for higher layers in the IoT architecture, enabling applications and systems to derive insights, make decisions, and take appropriate actions based on the captured information.

In general, the perception layer acts as the interface between the physical world and the digital realm of IoT, playing a crucial role in gathering data and initiating the flow of information within the IoT ecosystem.

Edge layer: The network layer in an IoT architecture plays a crucial role in facilitating the transmission of data between the perception layer and the application layer. Its primary function is to establish reliable and efficient communication channels to enable the seamless flow of information. Various communication technologies are utilized within the network layer to support IoT connectivity. These technologies include Wi-Fi, Bluetooth, Zigbee, and cellular networks, among others. Each technology offers different characteristics, such as range, data transfer rates, power consumption, and suitability for specific IoT use cases.

The network layer ensures that the data collected by the perception layer is transmitted securely and reliably to the application layer. It is responsible for handling tasks such as data aggregation, protocol conversion, and data routing. Additionally, it may employ techniques such as data compression and encryption to optimize data transmission efficiency and enhance security. One notable aspect of the network layer is its computational capability, which is typically greater than that of the perception layer. This allows for the implementation of edge computing on this layer. Edge computing involves processing and analyzing data closer to the source, reducing latency and enabling real-time or near-real-time responses. By performing computation tasks at the network layer, edge computing enhances the responsiveness and efficiency of time-sensitive IoT applications.

Furthermore, the network layer acts as a bridge between the IoT devices in the perception layer and the cloud or centralized servers in the application layer. It collects the IoT data, processes it if necessary, and facilitates its transmission to the cloud for further analysis, storage, and decision-making.

In summary, the network layer in an IoT architecture ensures secure and reliable data transmission between the perception and application layers. It leverages various communication technologies and incorporates edge computing capabilities to provide fast response times, particularly for time-sensitive applications.

Application layer: The application layer, also known as the cloud layer, represents the topmost layer of an IoT architecture. It encompasses the software applications and services that leverage the data collected from the perception layer to provide meaningful insights, perform analysis, and enable intelligent decision- making.

At the application layer, the focus shifts from raw data collection to utilizing the data to derive valuable information and take appropriate actions. This layer hosts a variety of applications, including data analytics, machine learning algorithms, and artificial intelligence (AI) models. These applications are responsible for processing and analyzing IoT data to extract patterns, trends, and actionable insights. One of the key functions of the application layer is to store and manage vast amounts of IoT data. The cloud serves as the central storage repository for IoT data, offering scalability, reliability, and accessibility. This centralized storage enables historical analysis and long-term data retention.

The application layer provides data access to IoT users, allowing them to inter- act with the IoT system and benefit from the insights generated. Users can access the data, monitor IoT devices, configure settings, and perform specific actions based on the information provided by the applications in this layer. This layer serves as a bridge between the underlying IoT infrastructure and the end-users or consumers of the IoT applications.

Due to its location in the cloud, the application layer benefits from significant computing power. This enables complex data processing tasks, advanced analytics, and AI-driven algorithms. The computational capabilities of this layer allow for sophisticated analysis, predictive modeling, and real-time decision-making. It's important to note that although the application layer possesses powerful computational capabilities, it is physically distant from the IoT devices in the perception layer. This physical separation necessitates efficient data transmission and communication protocols between the layers to ensure a timely and reliable exchange of information.

In summary, the application layer in an IoT architecture hosts software applications that analyze, process, and act upon the data collected from the perception layer. It leverages the cloud infrastructure, provides data access to users, and employs advanced analytics and AI techniques to deliver valuable insights and services to the end-users of IoT applications. This three-layered IoT architecture acts as a reference point for the literature review as well as while proposing the solution.

4.2.2 Overview of Blockchain technology

Blockchain is a distributed ledger technology that enables secure, transparent, and tamper-proof transactions between parties without the need for a central authority. The technology was first introduced by Nakamota (2008) as the underlying technology behind Bitcoin, a digital cryptocurrency.

The basic idea behind Blockchain technology is to create a decentralized database distributed across a network of computers. Each block in the chain contains a record of recent transactions, and once a block is added to the chain, it cannot be altered or deleted without consensus from the network. Liu (2019) mentioned that this makes the Blockchain an immutable and secure database. Blockchains can be public, where anyone can access the network and participate in the transaction validation process, or private, where access to the network is restricted to a select group of participants. Public Blockchains are often used for cryptocurrency transactions, while private Blockchains are used by enterprises for supply chain management, identity verification, and other applications. One of the key features of Blockchain technology is its use of cryptography to ensure the authenticity and integrity of transactions. Transactions are validated through a consensus mechanism, such as proof-of-work or proof-of-stake, which requires network participants to perform complex calculations to verify the transaction and add it to the Blockchain.

Many researchers, for example, Christidis and Devetsikiotis (2016), Beck (2018), and Lawand and Nitnaware (2023) have found that Blockchain technology has the potential to revolutionize the way transactions are conducted and recorded, providing a secure, transparent, and decentralized alternative to traditional centralized systems.

Secure aspect of blockchain technology: Blockchain technology works on the principle of a distributed ledger system that maintains a continuously growing list of records called blocks, which are cryptographically linked and secured using advanced cryptographic algorithms.

When a new transaction is initiated, it is broadcast to a network of nodes (also known as miners) that run specialized software to validate and verify the transaction. The miners then compete to solve a complex mathematical puzzle, and the first miner to solve the puzzle adds the new transaction to the existing Blockchain. This process is known as mining, and the miners are rewarded with a cryptocurrency for their efforts.

Once a block is added to the Blockchain, it cannot be altered or deleted, and all nodes in the network have a copy of the Blockchain, ensuring transparency and immutability of the transactions. To maintain the integrity of the Blockchain, it uses consensus algorithms such as Proof of Work (PoW), Proof of Stake (PoS), and Delegated Proof of Stake (DPoS) to ensure that all nodes in the network agree on the state of the Blockchain. This consensus mechanism ensures that the Blockchain is resistant to hacking attempts and ensures that the network remains secure and transparent.

4.2.3 Synergy between Blockchain and IoT

4.2.3.1 Related work

The convergence of Blockchain and IoT technologies has garnered substantial research attention due to its potential to address core challenges associated with security, data integrity, scalability, and interoperability. Several recent studies have proposed frameworks, protocols, and consensus mechanisms that serve as foundational pillars for the design and implementation of efficient Blockchain-based IoT systems.

Rangwala and Buyya (2024) introduced *TrustMesh*, a trusted distributed com- puting framework designed for heterogeneous IoT environments. Their architecture incorporates a permissioned Blockchain and a multiphase practical byzantine fault tolerance (PBFT) consensus protocol to facilitate secure and scalable interactions among decentralized devices. The layered architecture of TrustMesh aligns closely with the structured, modular approach proposed in this chapter, emphasizing fault tolerance and performance optimization.

Ramezan and Meamari (2024) proposed *zk-IoT*, a Blockchain-based architecture that enhances security through the use of zero-knowledge proofs (ZKPs). This approach ensures the integrity of firmware execution and data validation within potentially compromised IoT devices, offering an advanced trust model that supports autonomous interactions in trustless environments. Their work highlights the growing role of cryptographic advancements in reinforcing trust and privacy in IoT systems.

Guo, Li, and Nejad (2023) developed *LH-Raft*, a hierarchical and location-aware consensus protocol tailored to IoT–Blockchain integration. By grouping nodes based on geographical proximity and reputation, and leveraging threshold signatures for global consensus, their solution reduces communication overhead and latency. This protocol addresses scalability issues inherent in traditional Blockchain systems, particularly when deployed in large-scale, real-time IoT networks.

Haque, Shah, Iqbal, et al. (2024) proposed a scalable Blockchain-based data management framework utilizing DPoS and InterPlanetary File System (IPFS) for distributed data storage. The framework is specifically designed to operate efficiently within the constraints of IoT environments, demonstrating improvements in throughput, energy efficiency, and latency. Their contributions emphasize the value of lightweight consensus algorithms and modular storage systems for IoT scalability.

K. S. Kumar et al. (2024) explored a hybrid Blockchain-enabled architecture for Industry 4.0 applications, addressing the demands for secure data analytics and real-time communication in industrial IoT systems. Their work underscores the necessity of integrating Blockchain with real-time processing and analytics tools to meet the performance expectations of mission-critical environments.

Dinde and Shirgave (2024) provided a comprehensive review of Blockchain–IoT integration, categorizing prevailing security challenges and identifying future research opportunities. They emphasized the need for decentralized architectures, immutable device identities, and automated policy enforcement mechanisms. Their systematic evaluation reinforces the necessity of a generic, adaptable framework capable of supporting varied IoT application domains.

These contributions collectively underscore the rapid evolution of Blockchain-IoT integration strategies and validate the need for a unified, generic framework. The proposed work builds upon these foundations by introducing a comprehensive, modular, and adaptable development framework that incorporates best practices, emerging protocols, and implementation strategies to address core challenges in IoT–Blockchain integration.

4.2.3.2 Fundamental challenges

One of the fundamental challenges in IoT is the need for secure and verifiable authentication of devices in a dynamic and large-scale network. Traditional methods of authentication are often prone to vulnerabilities, as they rely on centralized authorities, which become single points of failure. Blockchain, with its decentralized consensus mechanism, removes the need for central authority intervention. Each IoT device can be registered and authenticated on the Blockchain, and the transaction history of each device can be made immutable, ensuring that only authorized devices can access the network. The use of cryptographic techniques, such as public–private key pairs and digital signatures, guarantees the authenticity of device interactions, mitigating risks related to identity spoofing and malicious attacks.

Data integrity is another significant concern in IoT systems, where the volume and velocity of data generated by a large number of devices can lead to in consistencies and potential tampering. Blockchain addresses this challenge by offering an immutable ledger of transactions. Each data point generated by an IoT device can be securely recorded on a Blockchain, ensuring that once recorded, the data cannot be altered or deleted. This feature of Blockchain guarantees the integrity of data transmitted across the IoT network, preventing unauthorized manipulation of sensor data or device readings. Additionally, consensus algorithms like PoW or PoS ensure that only verified data is included in the Blockchain, adding an additional layer of trust and security to the entire IoT ecosystem.

Trust is an overarching issue in IoT, as the proliferation of devices introduces a wide variety of stakeholders, each with potentially conflicting interests. In traditional centralized systems, trust is established through intermediaries, but these entities are susceptible to failures or attacks. Blockchain, with its distributed ledger architecture, inherently fosters trust among disparate parties by removing the need for intermediaries. The transparency and immutability of the Blockchain allow all participants to independently verify the data and transactions without the need for centralized oversight. Furthermore, Blockchain's integration with smart contracts facilitates automated decision-making processes based on predefined conditions, further enhancing trust in the system's operation and reducing the potential for human error or fraud.

In summary, the synergy between Blockchain and IoT is not merely an integration of two technologies but rather a transformative approach to the fundamental issues of security, data integrity, and trust in decentralized networks. Blockchain enables secure, transparent, and decentralized management of IoT data (Dange and P. Nitnaware 2023), addressing the scalability, authentication, and privacy concerns that have long hindered the widespread adoption of IoT in mission-critical applications.

4.3 REQUIREMENTS FOR A BLOCKCHAIN-BASED IOT FRAMEWORK

Developing a Blockchain-based IoT framework requires addressing several critical challenges to ensure a robust, efficient, and scalable system. Below are the key requirements for such a framework.

4.3.1 Scalability

The integration of IoT with Blockchain can result in a massive number of transactions due to the large volume of interconnected devices. To handle this, the framework needs to scale effectively. Techniques like "sharding", which partitions the Blockchain network to distribute the transaction load, and "layer-2 scaling solutions" (e.g., state channels, sidechains) are pivotal. These approaches aim to reduce the on-chain transaction volume, improving throughput without com- promising security. The scalability problem can be modeled by the transaction throughput T, expressed as:

$$T = n/t$$

where n is the number of transactions and t is the time taken. Techniques like **sharding** divide the network into S shards, theoretically increasing throughput to:

$$T_{\text{sharded}} = S \times n/t$$

Assuming perfectly parallel shards with no communication overhead or conflict.

4.3.2 Security and privacy

Security remains a fundamental concern in IoT networks due to the high risk of cyberattacks. Blockchain's cryptographic algorithms, such as "public-key cryptography" and "hash functions", provide a secure and tamper-proof mechanism for data transmission. Additionally, "smart contracts" help automate secure transactions between devices, mitigating risks of unauthorized access. Ensuring "data privacy" involves implementing privacy-preserving techniques like "zero-knowledge proofs" and "ring signatures", which allow for confidential transactions without exposing user data.

Blockchain security relies on cryptographic hash functions. Let $H(x)$ be a hash function outputting a fixed-length hash y. For privacy, zero-knowledge proofs allow a prover P to demonstrate knowledge of x such that:

$$H(x) = y$$

without revealing x. The probability of breaking this with a brute-force attack is:

$$P_{\text{attack}} = 1/2^k$$

where k is the hash length.

4.3.3 Interoperability

The heterogeneous nature of IoT devices and Blockchain platforms necessitates a framework that can facilitate seamless communication across diverse systems. Interoperability focuses on the compatibility between different protocols, standards, and Blockchain networks. This can be achieved through “cross-chain communication protocols” and ”middleware solutions” that enable interactions between distinct Blockchain ecosystems (e.g., Hyperledger Fabric, Ethereum) and various IoT devices, thus enhancing flexibility and integration capabilities.

Interoperability can be mathematically represented by mapping functions between protocol sets. For two protocols A and B, interoperability is defined if there exists a transformation function f such that:

$$f(A) \cong B$$

This ensures that data from A can be interpreted in B without loss.

4.3.4 Energy efficiency

IoT devices are often constrained by limited battery life and low computational power. Therefore, the Blockchain framework must prioritize energy-efficient solutions. Traditional consensus mechanisms like PoW are impractical due to high energy consumption. Instead, lightweight alternatives like PoS, DPoS, or Proof of Authority (PoA) are preferred. These methods significantly reduce energy usage, making them suitable for low-power IoT devices. The energy efficiency of consensus mechanisms can be modeled using energy cost, $E_{\text{consensus}}$. For N nodes using PoW, the energy consumption is:

$$E_{\text{PoW}} = N \times e_{\text{node}}$$

where e_{node} is the energy per node. Switching to PoS, the energy consumption E_{PoS} is:

$$\frac{E_{\text{PoS}}}{E_{\text{PoW}}} \ll 1$$

It indicates the energy consumption is very low in PoS.

4.3.5 Real-time processing

Real-time data processing is critical in IoT applications, especially for scenarios requiring immediate decision-making, such as in smart healthcare or autonomous vehicles. The framework must minimize latency to ensure swift data validation and transaction confirmation. Techniques like "edge computing" and "fog computing" can bring processing closer to the data source, reducing response time. Additionally, optimizing the consensus protocol for speed (e.g., using byzantine fault tolerance (BFT) mechanisms) enhances the framework's performance in real-time scenarios.

The latency L in a Blockchain IoT network can be minimized by edge computing.

$$L_{\text{edge}} \frac{1}{e_{\text{eff}}} \sum_{i=1}^{n} d_i$$

where $e_{\text{eff}} \leq e$ is the number of effective edge nodes, accounting for coordination, bandwidth, and local computation delays.

4.3.6 CUMULATIVE EFFICIENCY FORMULA

Combining these aspects, the overall efficiency E_{total} of the framework is

$$E_{total} = \alpha T_{sharded} * (1 - P_{attack}) * f_{interoperability} * \left[\frac{E_{Pow}}{E_{pos}} *\right]\left[\frac{1}{L_{edge}}\right]$$

α: Weighting factor for scalability.
$T_{sharded}$: Transaction throughput with sharding.
P_{attack}: Probability of a successful attack.
$f_{interoperability}$: Function assessing interoperability.
$\frac{E_{pow}}{E_{pos}}$: Ratio of energy efficiencies.
$\frac{1}{L_{edge}}$: Inverse of Latency

In summary, a successful Blockchain-based IoT framework must address these five requirements to effectively support the diverse and demanding needs of IoT applications. The higher throughput, lower attack probability, better interoperability and lower energy usage, and lower latency are required in Blockchain and IoT base system. The continuous evolution of Blockchain technologies and their integration with advanced IoT devices will be key to realizing scalable, secure, and efficient systems for future use cases.

4.4 FRAMEWORK ARCHITECTURE

The proposed architecture for a Blockchain-based IoT framework is designed to ensure scalability, security, and interoperability while meeting the dynamic requirements of IoT applications. This section outlines a comprehensive layered structure, a modular design approach, and integration interfaces, backed by existing research and new enhancements.

4.4.1 Layered structure

The architecture follows a layered model, which provides distinct responsibilities and enhances the scalability and maintainability of the system. This design is inspired by existing IoT and Blockchain integration studies, with adjustments to optimize performance across various IoT environments (Figure 4.2).

4.4.1.1 Device layer

This foundational layer comprises IoT sensors, actuators, and edge devices responsible for collecting and processing data. The integration utilizes communication protocols like MQTT (message queuing telemetry transport) and CoAP (constrained application protocol), known for their lightweight and efficient data transmission capabilities. These protocols ensure reliable connectivity even in resource-constrained environments. By employing edge computing, this layer reduces latency, facilitating immediate processing and decision-making close to the data source, thereby alleviating the computational burden on the network.

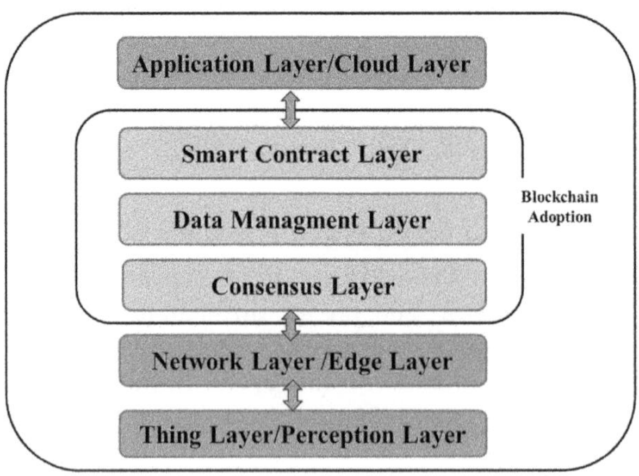

Figure 4.2 Proposed generic layered architecture.

4.4.1.2 Network layer

This layer handles data transmission through a peer-to-peer (P2P) network structure. It incorporates secure communication channels using transport layer security (TLS) and virtual private networks (VPNs) to prevent unauthorized access. Additionally, software-defined networking (SDN) is utilized for dynamic and efficient traffic management, optimizing data flow across IoT devices and the Blockchain network.

4.4.1.3 Consensus layer

The choice of consensus mechanism significantly impacts the framework's scalability and energy efficiency. For IoT scenarios, traditional mechanisms like PoW are inefficient due to high computational costs. Instead, lightweight alternatives such as PoA and DPoS are proposed. These mechanisms offer lower latency and energy consumption, enabling rapid consensus while maintaining security. Mathematically, the energy efficiency The energy efficiency, $E_{\text{consensus}}$, is modeled as:

$$E_{\text{consensus}} = \frac{1}{C}\sum_{i=1}^{N} E_i$$

where E_i is the energy used by node i, and C is the consensus complexity factor.

4.4.1.4 Data management layer

This layer focuses on data storage, balancing between on-chain and off-chain methods. On-chain storage offers immutability but is limited in capacity, making it suitable for critical, smaller data sets like transaction logs. Off-chain storage, through IPFS or cloud-based solutions, handles larger, non-critical data efficiently. The decision between on-chain D_{on} and off-chain D_{off} storage is optimized by:

$$D_{off=arg\ min}\left(\frac{S_{on}}{R_{on}}+\frac{S_{off}}{R_{off}}\right)$$

where S denotes data size and R the retrieval speed.

4.4.1.5 Smart contract layer

Smart contracts automate rule enforcement and decision-making in IoT operations. This layer supports creating decentralized applications (DApps) to handle specific tasks autonomously. The smart contract execution cost, defined by gas consumption G, is minimized through optimized code, calculated as:

$$G = \sum_{j=1}^{m} g_j . w_j$$

where g_j is the gas price of operation j and w_j its weight.

4.4.1.6 *Application layer*

The top layer is where user interaction takes place. It includes IoT applications such as real-time monitoring systems, dashboards, and control panels. This layer facilitates visual data analysis and integrates user commands for device control. The user experience is enhanced using responsive user interface (UI) frameworks and scalable backend services to handle variable data loads effectively.

4.4.2 Modular design

The modular design of a Blockchain-based IoT framework is a key architectural strategy aimed at enhancing flexibility, scalability, and customization. This approach segments the framework into independent, functional modules such as data management, smart contracts, consensus mechanisms, and integration interfaces. Each module can be independently updated or replaced without affecting the overall system, allowing seamless adaptation to new requirements or evolving technologies in IoT and Blockchain domains.

Key features

1. Scalability
   The modular approach allows each component to be scaled independently. For instance, the *consensus module* can be enhanced to accommodate more nodes without altering the data management layer. This modular flexibility is essential in IoT environments with fluctuating data loads and diverse application requirements.
2. Upgradability
   As Blockchain technologies evolve, the need for *upgrading consensus mechanisms* (e.g., transitioning from PoW to PoS) becomes apparent. With a modular structure, such upgrades are isolated to specific modules, reducing system downtime and mitigating risks.

3. **Customizability**
   Different IoT applications have varying requirements for data storage, security, and transaction processing. For example:

   - A healthcare IoT system may prioritize data privacy and require a specific encryption module.
   - A smart home system may focus on real-time responsiveness, necessitating a specialized consensus module.

The modular architecture accommodates these specific needs without a complete system overhaul.

Mathematical representation of modularity

The modularity M of the system can be expressed as:

$$M = \frac{\sum_{k=1}^{p} m_k}{T_{total}}$$

where

m_k: Performance or complexity metric of the k-th module (e.g., efficiency, processing speed).

p: Total number of modules in the system (e.g., data management, consensus, smart contract).

T_{total}: Overall system complexity or computational overhead.

Interpretation

A higher M value indicates a more modular system, which is better at integrating new technologies or adapting to diverse application requirements without significant changes.

Lower M values suggest tight coupling between components, limiting flexibility and increasing the complexity of updates or customizations.

Example: Consider a scenario where we have three primary modules:

Data Management ($m_1 = 0.8$),
Consensus Mechanism ($m_2 = 0.6$),
Smart Contract ($m_3 = 0.7$).

If the total system complexity T_{total} is 2.5, then the modularity is calculated as:

$$M = \frac{0.8 + 0.6 + 0.7}{2.5} = 0.84.$$

This value reflects good modularity, indicating that each module contributes independently to the system, enhancing the ability to adapt, upgrade, and scale according to specific IoT application requirements. Such a design framework, built on a modular principle, offers an adaptive and efficient solution suitable for diverse and evolving IoT ecosystems.

4.4.3 Integration through middleware and APIs

To facilitate seamless communication between IoT devices, Blockchain networks, and external systems, the proposed framework employs application programming interfaces (APIs) and middleware.

4.4.3.1 Role of middleware

1. *Protocol translation*: Middleware acts as a bridge, converting data formats and ensuring compatibility between disparate communication protocols (e.g., MQTT to HTTP). This translation allows devices using different protocols to interact smoothly with the Blockchain network.
2. *Data Formatting and standardization*: It ensures that data from various IoT devices is formatted consistently, enabling the Blockchain network to process it without errors or mismatches.
3. *Secure data exchange*: Middleware implements encryption and secure channels (e.g., TLS/SSL) to protect the data during transmission, ensuring that communication remains confidential and tamper-proof.

Mathematical expression of integration

The integration function I represents the seamless interaction between a device and the Blockchain network. It is defined as:

$$I = f(A, B, C)$$

where

A: Compatibility of Device Protocols – Evaluates if the device's communication protocol (e.g., MQTT, CoAP) is compatible with the network requirements.

B: Network Security Standards – Assesses the security measures, including encryption and authentication, needed for secure communication be- tween the device and the Blockchain.

C: Data Handling Capabilities – Represents the ability of the system to handle data in terms of size, format, and processing requirements.

Example
For instance, if a temperature sensor (using CoAP) wants to send data to a Blockchain network designed for smart contracts, middleware translates the CoAP protocol into HTTP/JSON format, formats the sensor data to align with Blockchain specifications, and encrypts it before transmission. This integration model ensures interoperability across different platforms and devices, security through encrypted data exchange and consistency by formatting data to meet Blockchain standards.

Such an approach enhances the framework's flexibility and reliability, accommodating a diverse range of IoT applications and devices.

4.5 IMPLEMENTATION APPROACH/STRATEGY

To facilitate the integration of Blockchain into IoT applications using the proposed layered framework, the implementation strategy is outlined in a step-by-step approach, ensuring clarity for both technical and non-technical audiences. This strategy reflects the layered structure while maintaining high scientific rigor.

4.5.1 Device layer: Building the IoT infrastructure

The device layer forms the foundation for any IoT system, as it enables the collection of data from the physical environment and ensures interaction with various IoT devices. This layer involves the integration of sensors, actuators, and edge de- vices that communicate with each other and relay data to the Blockchain network.

Objective
The primary goal of the device layer is to gather data from the environment through IoT sensors and facilitate interaction between the IoT devices and the Blockchain network. It acts as the entry point for data that is later processed and validated by higher layers.

Components and hardware
To build this layer, hardware like Raspberry Pi, Arduino, or other edge devices are employed to simulate real-world sensors and actuators. These devices are equipped with various types of sensors (e.g., temperature, humidity, and pressure) and actuators that can influence the environment based on the collected data.

Steps for implementation

1. *Configure IoT sensors*: Set up IoT sensors (such as temperature, humidity, or motion sensors) to monitor the environment. These

sensors gather real-time data that will be transmitted for further processing.

2. *Use lightweight communication protocols*: Implement communication protocols like MQTT or COAP for transmitting sensor data. These protocols are optimized for low-power and low-bandwidth environments, a typical scenario in IoT systems.
3. *Implement Edge computing*: Edge computing refers to processing data locally on the device or nearby edge nodes rather than relying solely on cloud computing. This reduces latency, enhances response times, and alleviates the need for high-volume data transmission to the central server.
4. *Secure data transmission*: Ensure that the data transmitted from IoT devices is encrypted and securely sent to the next layer, typically using secure protocols like TLS/SSL or other security measures to prevent tampering or data breaches.
5. *Data preprocessing (edge layer)*: Preprocess the collected data locally before transmission. This involves filtering, cleaning, and possibly aggregating the data to reduce the amount of data sent to the Blockchain network. For instance, data can be aggregated every few minutes rather than sending continuous streams.

Outcome: The outcome of the device layer implementation is the establishment of a robust system for data collection, transmission, and preprocessing. The layer enables IoT devices to generate structured data that can be reliably transmitted to the Blockchain network for further processing and validation. This design ensures IoT network operates efficiently and that communication between devices and the Blockchain is optimized for performance and security. The preprocessing at the edge further reduces the load on the Blockchain network, ensuring that only relevant data is transmitted.

Network layer: Ensuring secure data transmission

The network layer serves as the backbone for communication between IoT devices and the Blockchain network. This layer establishes a secure and efficient communication channel, ensuring data integrity, confidentiality, and reliable transmission.

Objective

The primary goal of the network layer is to facilitate secure, stable, and scalable data exchange between IoT devices and the Blockchain network. It ensures that transmitted data is encrypted and resistant to unauthorized access or tampering.

Components
The key components of the network layer include:

Blockchain networks: Use private or hybrid Blockchain platforms like Hyperledger Fabric or Ethereum (PoA) to ensure efficient data processing and consensus mechanisms.
Communication protocols: Employ peer-to-peer (P2P) protocols to enable direct communication between nodes.
Security measures: Utilize cryptographic techniques such as TLS/SSL to encrypt data during transmission.

Steps for implementation

1. *Establish peer-to-peer communication*
   Implement P2P communication to ensure direct and decentralized data exchange between IoT devices and Blockchain nodes. This minimizes the risk of data interception and enhances network efficiency.
   Configure network topology to ensure redundancy and fault tolerance, preventing data loss in case of node failure.
2. *Implement secure communication channels*
   Use **TLS/SSL** protocols to create secure communication links, encrypting all data transmitted over the network.
   Integrate advanced encryption techniques like AES-256 for data packets to ensure confidentiality and prevent unauthorized access.
3. *Test network stability*
   Simulate variable conditions such as fluctuating data loads, varying node availability, and network disruptions.
   Evaluate metrics like throughput, latency, and packet loss to ensure scalability and reliability under different scenarios.
4. *Integrate network monitoring tools*
   Deploy monitoring tools to track network performance and detect anomalies in real time.
   Implement logging mechanisms for detailed analysis and debugging of network issues.

Outcome: By implementing the network layer, the following can be achieved:

A secure and encrypted communication channel is established between IoT devices and the Blockchain.
Stable and scalable network performance is achieved, even under high data loads and varying conditions.

The layer ensures that data integrity and confidentiality are maintained throughout the transmission process.

This robust and secure network layer is critical for ensuring the seamless integration of IoT devices with the Blockchain network, enabling reliable data transmission and supporting the overall system's functionality.

Consensus layer: Adopting lightweight mechanisms

The consensus layer is a critical component that ensures agreement among participating nodes in the Blockchain network while maintaining the integrity and security of the data. For IoT applications, lightweight consensus protocols are essential to address constraints such as limited processing power, intermittent connectivity, and energy efficiency.

Objective

The primary goal of the consensus layer is to implement efficient and scalable consensus mechanisms tailored for IoT environments. These mechanisms ensure data consistency across the network while minimizing computational overhead and energy consumption.

Components

The key components of the consensus layer include:

Consensus protocols: Lightweight mechanisms such as PoA or DPoS are preferred over energy- intensive protocols like PoW.

Fault tolerance measures: Techniques to ensure network reliability even during node failures or intermittent IoT device connectivity.

Simulation tools: Platforms to simulate and analyze consensus performance under various IoT-specific scenarios.

Steps for implementation

1. *Deploy lightweight consensus algorithms*

   Implement PoA): A mechanism where a limited number of trusted validators authenticate transactions, reducing computational and energy requirements.

   Utilize DPoS: Allows token holders to vote for delegates who are responsible for transaction validation, ensuring high throughput.
2. *Ensure resilience to IoT-specific challenges*

   Address issues like intermittent connectivity by incorporating fault-tolerant designs.

   Use techniques such as asynchronous validation to handle delayed transactions without disrupting network operations.

3. *Simulate and test performance*
   Conduct simulations under various loads to evaluate metrics such as transaction throughput, latency, and network scalability.
   Analyze how the consensus protocol responds to high node churn rates, ensuring reliability in dynamic IoT environments.

Outcome: By implementing the consensus layer, the following can be achieved:

Reduced energy consumption: Lightweight mechanisms like PoA and DPoS minimize energy usage, making them suitable for resource-constrained IoT devices.
Enhanced processing speed: Faster consensus mechanisms ensure high throughput and real-time processing for IoT applications.
Scalable and secure network: The layer supports the addition of new devices while maintaining data integrity and security, even under challenging network conditions.

This consensus layer forms the backbone of a secure, efficient, and scalable Blockchain network for IoT, addressing the unique constraints of IoT ecosystems while maintaining robust operational performance.

Data management layer: Handling and storing IoT data
The data management layer is designed to manage, store, and retrieve data generated by IoT devices efficiently. Given the high volume of IoT data, a hybrid data storage model is essential to optimize Blockchain utilization while balancing cost and performance.

Objective
The objective is to create a hybrid storage strategy that uses Blockchain for critical data and off-chain solutions for bulk data, ensuring security, scalability, and cost-effectiveness.

Components
The key components of the data management layer include:

On-chain storage: Utilized for storing critical data summaries or metadata securely.
Off-chain storage: Systems like IPFS or cloud platforms are used for storing large volumes of raw IoT data.
Hashing techniques: Ensures data integrity by creating unique digital fingerprints of data stored off-chain.

Steps for implementation

1. *Secure on-chain storage with hashing*
   Generate a hash of the IoT data using cryptographic algorithms (e.g., SHA-256) to create a secure, tamper-proof summary.
   Store the hash on the Blockchain to ensure traceability and data integrity.
2. *Utilize decentralized and cloud-based storage for bulk data*
   Leverage IPFS: A decentralized file storage and sharing protocol to store bulk data efficiently.
   For applications requiring scalability, integrate cloud platforms such as AWS, Azure, or Google Cloud.
3. *Link on-chain and off-chain data*
   Associate the on-chain hash with the off-chain storage location, enabling consistent and verifiable data retrieval.
   Use smart contracts to automate the linking process, ensuring real-time updates and synchronization.
4. *Ensure data consistency and retrievability*
   Periodically verify that the off-chain data matches the on-chain hash.
   Implement redundancy mechanisms to ensure data availability and re duce the risk of data loss.

Outcome: The data management layer results in the following:

Efficient storage: By storing only critical data on-chain and using off-chain storage for bulk data, the model reduces Blockchain bloat.
Cost optimization: Minimizes Blockchain storage costs while maintaining data integrity.
Scalable retrieval mechanism: Ensures seamless data retrieval through well-defined links between on-chain and off-chain storage.
Enhanced security: Hashing techniques provide tamper-proof evidence of data integrity.

This approach optimizes data management for IoT applications, ensuring a balance between security, performance, and cost-effectiveness.

Smart contract layer: Automating IoT operations
The smart contract layer is crucial for automating tasks and validating data within IoT environments. By using Blockchain technology, smart contracts help to streamline operations, improve efficiency, and ensure security. This layer enables IoT devices to perform actions autonomously, based on predefined conditions, without the need for central authority intervention.

Objective

The primary goal of the smart contract layer is to automate IoT tasks such as data validation, triggering alerts, and executing automated actions based on certain conditions. This is achieved by developing smart contracts that interface with IoT devices, ensuring seamless operation within the system.

Components

The key components of the smart contract layer include:

Smart contracts: Self-executing contracts with the terms of the agreement between buyer and seller directly written into lines of code. These automate the execution of predefined tasks.

Tools and frameworks: Smart contracts for Ethereum are typically written in Solidity, while for Hyperledger Fabric, Chaincode is used. These frameworks provide the necessary environment to develop and deploy smart contracts.

IoT integration: Smart contracts are designed to interact with IoT sensors and devices, triggering actions based on sensor readings or external inputs.

Steps for implementation

1. *Smart contract development*
   Develop smart contracts using Solidity (for Ethereum) or Chaincode (for Hyperledger Fabric). These contracts will define actions like data validation, monitoring thresholds, or triggering alerts.
   Write functions within the smart contract to verify sensor data, ensuring that it is within acceptable ranges before allowing further actions (e.g., triggering an alert or initiating a response).
2. *Automate IoT tasks*
   Create automated workflows based on sensor readings or events. For instance, when a temperature sensor detects a threshold breach, the smart contract can automatically trigger a cooling system or send an alert.
   Use Oracles to fetch real-world data (such as weather information) that may influence the execution of smart contracts.
3. *Testing and security auditing*
   Conduct rigorous testing, including unit testing to verify individual contract functions, and integration testing to ensure smooth interaction with IoT devices and other system components.
   Perform security audits to identify and address potential vulnerabilities in the contract code, ensuring that the contract executes safely and as intended without risks of exploitation.

4. *Deployment and monitoring*
   Deploy the smart contract onto the Blockchain (e.g., Ethereum or Hyperledger Fabric) after successful testing.
   Continuously monitor the performance and behavior of deployed smart contracts to ensure that they function correctly, providing automated actions in response to IoT device data.

Outcome: The smart contract layer ensures the following outcomes:

Automated IoT task management: Tasks such as data validation, threshold-based actions, and alerting are automatically handled by the smart contract, reducing the need for manual intervention.

Enhanced security and transparency: Blockchain's immutable nature ensures that the executed contracts are transparent and tamper-proof, providing security against unauthorized changes.

Scalability and efficiency: Smart contracts execute quickly and reliably, improving the scalability of IoT applications and making them more efficient.

Cost savings: By automating processes and removing the need for intermediary parties, smart contracts help reduce operational costs in IoT systems.

In conclusion, the smart contract layer plays a pivotal role in automating IoT operations, improving efficiency, and ensuring the security and integrity of the system.

Application layer: Building user-friendly interfaces

The application layer serves as the user interface for interacting with the IoT system and Blockchain network. It provides a way for users to visualize, monitor, and control IoT devices, while also offering insights into Blockchain transactions and system activities. A well-designed application layer enhances usability and makes complex IoT–Blockchain ecosystems accessible to a broad range of users.

Objective

The objective of the application layer is to design intuitive and interactive interfaces that allow users to easily manage and interact with the IoT system. This includes visualizing real-time data, controlling devices, and monitoring Blockchain activities in an efficient manner.

Components

The key components of the application layer include:

User dashboards: Visual interfaces where users can view real-time IoT data, such as sensor readings, device status, and Blockchain-related activities (e.g., transaction logs, smart contract interactions).

Control panels: Interactive sections that allow users to manage IoT devices, configure system settings, and execute actions, such as triggering smart contracts or controlling IoT devices.

Cross-platform compatibility: The application should be accessible across different platforms, including desktop, mobile, and tablet devices, ensuring a responsive and consistent user experience.

Blockchain activity visualization: Displays logs, transaction history, and contract execution details to ensure transparency and allow users to track the flow of data within the Blockchain network.

Steps for implementation

1. *Dashboard design*

   Design dynamic dashboards that display real-time IoT data, including sensor readings (e.g., temperature, humidity) and IoT device statuses.

   Integrate Blockchain activity visualization, such as transaction logs, smart contract execution details, and Blockchain health monitoring (e.g., network nodes, mining activities).

2. *Control panel development*

   Develop control panels that allow users to interact with IoT devices. This could include toggling device states, adjusting settings (e.g., thresholds for sensor data), and executing actions based on IoT data or Blockchain events.

   Ensure the ability to manage smart contracts, allowing users to trigger, monitor, and audit contract executions related to IoT operations.

3. *Usability and accessibility*

   Use responsive design principles to ensure that the application works seamlessly across various screen sizes and devices, including smartphones, tablets, and desktops.

   Focus on user experience (UX) by ensuring that the application is easy to navigate, intuitive to use, and visually appealing.

4. *Integration with blockchain and IoT*

   Integrate the application with IoT device APIs to collect real-time data and transmit commands to IoT devices.

   Connect to the Blockchain network (e.g., Ethereum or Hyperledger) to fetch transaction details, smart contract executions, and other Blockchain events.

5. *Security and authentication*
   Implement secure user authentication mechanisms (e.g., multifactor authentication, role-based access control) to protect sensitive data and interactions.
   Ensure secure communication between the application, IoT devices, and the Blockchain network using encryption techniques (e.g., SSL/TLS).

Outcome: The successful implementation of the application layer results in the following:

User-friendly interfaces: Easy-to-use dashboards and control panels that allow users to efficiently manage IoT devices and monitor Blockchain activities.
Real-time visualization: Instant access to real-time sensor data and Blockchain transaction logs, offering transparency and insights into system operations.
Cross-platform accessibility: A responsive application that functions seamlessly across all major devices and operating systems, enhancing the user experience.
Enhanced security: Robust authentication and encryption mechanisms that ensure secure user interactions with the IoT and Blockchain systems.

In summary, the application layer is integral to providing an intuitive and interactive interface for users to engage with the IoT–Blockchain ecosystem. By focusing on usability, real-time visualization, and cross-platform compatibility, this layer ensures that the system is accessible to both technical and non-technical users.

Testing and deployment: Ensuring real-world functionality
The testing and deployment phase is critical for ensuring that the proposed IoT– Blockchain system performs effectively under real-world conditions. This stage involves validating the system's functionality through various testing methodologies and deploying it in actual IoT environments to measure its real-world performance. The goal is to ensure that the system can handle real-time data, scale efficiently, and maintain security and reliability when integrated with Blockchain technology.

Objective
The objective of this phase is to rigorously test the IoT–Blockchain framework to validate its functionality, scalability, and security, followed by deployment in real-world environments. By performing extensive testing,

we aim to identify and resolve potential issues before the system is fully deployed.

Steps for implementation

1. *Comprehensive testing*
   Unit testing: Each individual component (e.g., IoT sensors, smart contracts, Blockchain transactions) is tested independently to ensure that it functions as expected in isolation.
   Integration testing: After unit testing, different components are integrated and tested together. This ensures that the system components (IoT devices, Blockchain, and smart contracts) work cohesively when interacting with each other.
   Security testing: Since the system involves Blockchain and IoT devices, it is essential to perform thorough security testing. This includes testing for vulnerabilities such as unauthorized access, data breaches, and attacks like Sybil or DDoS attacks. Techniques such as penetration testing and threat modeling are used to identify potential security risks.
2. *Deploying in controlled IoT environment*
   Prototype deployment: Initially, deploy the system in a controlled environment (e.g., a laboratory or simulation setup) to evaluate its performance in terms of latency, throughput, and reliability.
   Monitoring metrics: During the prototype phase, monitor key performance metrics like transaction throughput (TPS), latency, energy consumption, and error rates to identify any system weaknesses or bottlenecks.
3. *Scaling deployment to real-world IoT applications:*
   industry-specific deployment: Once the prototype performs satisfactorily, scale the deployment to real-world IoT applications, such as smart cities, healthcare systems, or industrial automation. These use cases will help assess the system's effectiveness in different environments.
   Scalability testing: As the system scales, ensure it maintains performance across a diverse set of devices, varying data loads, and unpredictable network conditions. Evaluate the system's ability to handle increased data volume and device interactions.
   Integration with the existing infrastructure: Verify that the framework can integrate seamlessly with existing IoT infrastructure, whether legacy systems or modern devices, ensuring compatibility with various IoT protocols, standards, and data formats.

4. *Continuous monitoring and optimization*
   Post-deployment monitoring: Once the system is deployed in the field, establish continuous monitoring to track its performance and health. Utilize monitoring tools to assess device status, Blockchain activity, transaction integrity, and overall system health.
   Feedback loop: Based on real-world performance and user feedback, refine and optimize the system. This includes bug fixes, software up dates, and adjustments to improve efficiency, scalability, and security.
   Outcome: The outcome of the testing and deployment phase is a robust, fully functional IoT system integrated with Blockchain technology. After extensive testing, the system has the following characteristics:
   Validated: It has been thoroughly tested under various scenarios and conditions to ensure that all components function correctly and securely.
   Scalable: The system is capable of scaling across different IoT applications, from small-scale pilot deployments to large-scale production environments, while maintaining high performance.
   Resilient: The system is resilient to security threats and is capable of operating under varying network conditions and device interactions without failure.
   Ready for production: The system is fully deployed and optimized for use in real-world environments, ensuring it meets specific industry requirements.

In summary, the testing and deployment phase ensures that the IoT–Blockchain framework is validated for performance, security, and scalability, making it ready for real-world application. By conducting comprehensive testing and deploying in controlled and real-world environments, we ensure that the system can perform reliably in diverse use cases and applications.

4.6 CASE STUDIES

Case study 1: Smart Healthcare Monitoring System

Introduction

The Smart Healthcare Monitoring System is designed to continuously monitor patients' vital signs using wearable IoT devices. This system integrates with Blockchain technology to ensure data integrity, automate responses based on preset thresholds, and maintain security and decentralization. Blockchain helps in achieving immutable patient records and automates workflows such as alerting healthcare providers when abnormal data is detected.

System architecture and implementation

IoT device layer: Wearable sensors, such as heart rate monitors, temperature sensors, and blood pressure cuffs, continuously collect health data from patients. These devices transmit the collected data to edge nodes for local preprocessing. The data is then sent securely to the Blockchain network for logging and processing.

Blockchain integration: The patient data is hashed and logged onto the Ethereum Blockchain using a PoA consensus mechanism to enhance efficiency. Smart contracts are deployed to automate actions like triggering alerts when health metrics exceed a predefined threshold.

Data storage: Critical health data, such as heart rate and blood pressure, is stored on the Blockchain to ensure integrity and traceability. Non-critical data, such as full medical records, is stored off-chain using decentralized storage systems like IPFS, reducing the on-chain storage burden.

Benefits of Blockchain

1. Data integrity (immutability of health data)

The health data for each patient can be represented as a dataset $D = \{d_1, d_2, \ldots, d_n\}$, where each data point d_i is hashed and stored in a Blockchain. Each block B_{i+1} is linked to the previous block B_i via the hash of the previous block:

$$B_{i+1} = H(B_i) \cup d_{i+1}$$

This ensures that any alteration in data d_i would result in a change in the hash, disrupting the chain. Therefore, the integrity of patient health data is maintained with high probability, i.e., $P_{\text{integrity}} = 1$.

2. Automation of health alerts (smart contract execution)

Smart contracts are used to automate actions. For instance, if a patient's heart rate hr exceeds a threshold T, the smart contract triggers an alert:

$$f(hr) = \begin{cases} \text{Alert, if } hr > T \\ \text{No action, if } hr \leq T \end{cases}$$

The number of transactions handled by the smart contract per day is given by:

$$N_{\text{transactions}} = m \times (\text{average checks per patient per day})$$

where m is the number of patients. This reduces the need for manual interventions and ensures quick, automated responses.

1. Data security (encryption and access control)

Health data is encrypted using a cryptographic function E to ensure that it is accessible only to authorized entities A. For each patient's health data H_i, encryption is performed:

$$E(H_i) = \text{Encrypted Data, where } A \in \{\text{authorized entities}\}$$

The security is governed by the strength of the encryption key, and the probability of a successful attack is:

$$P_{\text{attack}} = \frac{1}{2^L}$$

where $L = 256$ for AES encryption, ensuring that the attack probability is negligible, thus protecting patient confidentiality.

2. Decentralization (data availability)

The decentralized nature of Blockchain ensures that patient data remains accessible even if some nodes fail. Suppose the Blockchain network consists of N nodes, each maintaining a full copy of the ledger. Assuming independent node failures with probability q, the system availability A_{system} defined as the probability that at least one node remains operational can be expressed as:

$$\text{Asystem} = 1 - q^N$$

For example, if each node has a 10% chance of being unavailable ($q = 0.1$) and the network consists of $N = 10$ nodes, then:

$$A_{\text{system}} = 1 - (0.1)^{10} = 1 - 10^{-10} \approx 1$$

This illustrates the robustness of decentralized systems, where data remains accessible despite multiple node failures, thus ensuring high availability.

Outcome: The proposed Blockchain-based IoT framework ensures

1. *Data integrity*: Patient health data is tamper-proof, fostering trust.
2. *Automation*: Smart contracts reduce manual work and automate health alerts.
3. *Security*: Patient data is securely encrypted and access-controlled.
4. *Decentralization*: Blockchain ensures continuous access to data even in case of node failures.

This combination of IoT and Blockchain results in a reliable, secure, and efficient healthcare monitoring system that meets both current and future needs.

Case study 2: Smart Crop Management System

Introduction

The Smart Crop Management System leverages IoT devices to monitor and optimize agricultural practices. Integrating Blockchain technology ensures transparent, secure, and tamper-proof data for farming operations. The proposed system can help farmers make data-driven decisions related to irrigation, pest control, and crop health monitoring, enhancing yield and resource efficiency. Blockchain technology also facilitates decentralized control and secure transactions in agricultural supply chains.

System architecture and implementation

IoT device layer: IoT sensors placed in the field collect real-time data related to soil moisture, temperature, humidity, and pest activity. Drones and automated systems can further monitor crop health and growth. The data is transmitted securely to edge devices for preprocessing and sent to the Blockchain network for logging and processing.

Blockchain integration: The data from various sensors is hashed and logged onto a Blockchain platform such as Hyperledger Fabric, chosen for its ability to handle high throughput and scalability. Smart contracts are implemented to automate processes such as irrigation control based on soil moisture readings and to trigger actions when pests are detected.

Data storage: Critical sensor data (soil moisture levels, pest detection) is stored on the Blockchain, ensuring traceability and tamper-proof records. Large datasets (e.g., historical crop performance, weather data) are stored off-chain using decentralized storage systems like IPFS, maintaining cost-effectiveness.

Blockchain benefit

1. Data integrity (immutable crop data logs)

Every piece of data, such as soil moisture readings M_i or pest count P_i, is hashed and logged onto the Blockchain. The integrity of the data is assured because any modification in the data would invalidate the Blockchain chain:

$$B_{i+1} = H(B_i) \cup M_i \cup P_i$$

This guarantees that all recorded data cannot be altered without detection. The probability of data manipulation in a Blockchain with n blocks is given by:

Which is negligible for large n, ensuring data integrity and trustworthiness.

$$P_{\text{manipulation}} = \frac{1}{2^n}$$

2. Automation of irrigation and pest control (smart contract execution)

Smart contracts are used to automate the irrigation process. For example, if soil moisture M falls below a threshold $T_{\text{irrigation}}$, an automated irrigation system is triggered:

$$f(M) = \begin{cases} \text{Activate irrigation, if } M < T_{irrigation} \\ \text{No action, if } M \geq T_{irrigation} \end{cases}$$

Similarly, if the pest count P exceeds a threshold T_{pest}, the smart contract triggers a pesticide application:

$$f(P) = \begin{cases} \text{Activate pesticide application, if } P > T_{pest} \\ \text{No action, if } P \leq T_{pest} \end{cases}$$

The number of transactions executed by the smart contract per day depends on the frequency of data checks and the number of sensors, represented by:

$$N_{\text{transactions}} = m \times n \times (\text{check frequency per sensor per day})$$

where m is the number of fields and n is the number of IoT devices deployed per field.

3. Supply chain transparency and security (secure transactions)

Blockchain facilitates secure, transparent transactions between farmers, suppliers, and buyers. The transactions involving crop yields, pesticide purchases, and equipment rentals are securely recorded on the Blockchain, reducing fraud and increasing trust. If the transaction value is V and the Blockchain uses a crypto- graphic algorithm with key length L, the security level can be quantified by:

$$p_{\text{fraud}} = \frac{1}{2^L}$$

where $L = 256$, ensuring high resistance to fraud and unauthorized transactions.

4. Decentralization (resilient and transparent system)

The decentralized nature of Blockchain ensures that no single entity controls the entire system, thus preventing manipulation and enhancing transparency. In such a system, each of the N nodes maintains a full copy of the Blockchain, contributing to overall data availability and system resilience. Assuming each node has an independent probability q of failure, the overall system availability A_{system} defined as the probability that at least one node remains operational can be expressed as:

$$Asystem = 1 - q^N$$

For instance, if each node has a 10% chance of failure ($q = 0.1$) and the network has $N = 10$ nodes, the system availability becomes:

$$A_{\text{system}} = 1 - (0.1)^{10} = 1 - 10^{-10} \approx 1$$

This high availability demonstrates the robustness of decentralized Blockchain systems, ensuring that data and transactions remain accessible even during multiple node failures.

Outcome: The proposed IoT–Blockchain framework for crop management ensures the following:

Data integrity: Sensor data related to soil conditions and pest activity is immutable, providing reliable records for farming operations.

Automation: Smart contracts automate irrigation and pest control actions, reducing manual intervention and resource wastage.

Security: Blockchain secures financial transactions within the agricultural supply chain, preventing fraud.

Decentralization: A decentralized system reduces the risks associated with single points of failure, ensuring continuous operation.

This integration of IoT and Blockchain ensures improved crop yield, optimized resource use, and enhanced trust in agricultural processes.

4.7 CHALLENGES AND SOLUTIONS

1. *Scalability issues*

Challenge: Blockchain networks often face scalability limitations when handling high transaction volumes from IoT devices. The time required for block confirmation T_c increases with the number of transactions N_t and nodes N_n:

$$T_c = f\,(N_t, N_n)$$

where f represents the network's scalability function.

Solution: Off-chain processing techniques, such as state channels or sidechains, can alleviate this issue by reducing the on-chain transaction load:

$$\text{Non-chain} = Nt - \text{Noff-chain}$$

where $n_{\text{on-chain}}$ represents transactions committed to the main Blockchain, and $n_{\text{off-chain}}$ are transactions handled externally. Lightweight consensus mechanisms like sDPoS or PoA further optimize T_c by minimizing computational overhead.

2. *Security and privacy risks*

Challenge: IoT ecosystems are susceptible to cyberattacks, such as data breaches and unauthorized access. The probability of a successful attack P_{attack} depends on the strength of cryptographic protocols C_s and the number of attack attempts A_t:

$$P_{\text{attack}} = \frac{A_t}{2^{C_s}}$$

Solution: Blockchain enhances security by using cryptographic hashes and distributed consensus. For data privacy, encryption techniques and anonymization methods can be integrated into the framework:

$$H(D) = \text{SHA } 256(D), \quad D_{\text{encrypted}} = E_k(D)$$

where $H(D)$ is the hash of data D, and $E_k(D)$ is the encryption of D using key k. Distributed ledgers further ensure that data remains tamper-proof.

3. *Regulatory and compliance challenges*

Challenge: Integrating Blockchain with IoT systems must comply with industry- specific regulations and data sovereignty laws. The complexity C_r of compliance grows with the number of regulations N_r and jurisdictions J:

$$C_r = f(N_r, J)$$

Solution: Blockchain frameworks can incorporate rule-based smart contracts to enforce compliance automatically:

$$\text{SC}_{(\text{Ri})} = \begin{cases} \text{Compliant, if Ri is satisfied} \\ \text{Non-compliant, otherwise} \end{cases}$$

where R_i represents individual regulatory requirements. Auditable Blockchain records ensure traceability and facilitate compliance reporting.

4. *Energy consumption*

Challenge: IoT devices often operate in resource-constrained environments, making energy efficiency crucial. The energy consumed E in a Blockchain system depends on the number of transactions N_t, consensus mechanism C_m, and computational complexity $O(n)$:

$$E = k \cdot N_t \cdot O(n)$$

where k is the energy coefficient.

Solution: Energy-efficient consensus mechanisms like PoA reduce computational complexity. Additionally, edge computing minimizes the energy required for data transmission by preprocessing data locally:

$$E_{edge} = E_{total} + E_{transmission}$$

This approach balances energy use without compromising system performance.

5. *Adoption barriers*

Challenge: Resistance to adopting Blockchain-based IoT systems arises from factors such as cost, complexity, and lack of awareness. Adoption rate A_r is inversely proportional to perceived complexity C_p and cost C_c:

$$A_r = \frac{1}{C_p + C_c}$$

Solution: The proposed framework addresses these barriers by:

Simplifying system deployment using modular architecture, reducing C_p.
Demonstrating cost-effectiveness through hybrid storage models, minimizing C_c.
Educating stakeholders on Blockchain benefits using transparent case studies and pilot projects.

By addressing these challenges with scientifically grounded solutions, the proposed framework ensures scalability, security, regulatory compliance, energy efficiency, and ease of adoption. These measures collectively support the successful integration of Blockchain technology into IoT ecosystems, making them more robust and sustainable.

4.8 CONCLUSION

This chapter explored the integration of Blockchain technology with IoT applications, proposing a robust, layered framework designed to address key challenges in scalability, security, and data management. Each layer of the framework was meticulously detailed, from the IoT device layer to the application layer, with an emphasis on efficient data collection, secure transmission, lightweight consensus mechanisms, hybrid storage solutions, and user-friendly interfaces. Implementation strategies and case studies further demonstrated the practical applicability of the framework across diverse domains.

The convergence of Blockchain and IoT presents a transformative opportunity to revolutionize industries by enhancing transparency, security, and efficiency. The proposed framework not only addresses the limitations of current systems but also paves the way for innovative applications in sectors such as healthcare, agriculture, and smart cities. By incorporating advanced technologies like edge computing and AI, this framework offers a scalable, future-proof solution to the growing demands of IoT ecosystems.

We urge researchers, developers, and industries to adopt and build upon this proposed framework, leveraging its modular architecture to create customized solutions for specific IoT applications. Future explorations should focus on refining Blockchain protocols, improving IoT device interoperability, and addressing emerging challenges in regulatory compliance and energy efficiency. Collaborative efforts in these areas will be instrumental in unlocking the full potential of Blockchain-enabled IoT systems and driving innovation across the global landscape.

REFERENCES

Beck, Roman (2018). "Beyond Bitcoin: The Rise of BBlockchain World". In: *Computer* 51.2, pp. 54–58.

Christidis, Konstantinos and Michael Devetsikiotis (2016). "Blockchains and Smart Contracts for the Internet of Things". In: *IEEE Access* 4, pp. 2292–2303. DOI: 10.1109/ ACCESS.2016.2566339.

Dange, Smita and Madhumita Chatterjee (2020). "A Broad Perspective on Integrating Internet of Things with Blockchain". In: *2020 5th International Conference on Computing, Communication and Security (ICCCS)*, pp. 1–5. DOI: 10.1109/ ICCCS49678.2020.9277062.

Dange, Smita and Prashant Nitnaware (2023). "Secure Share: Optimal Blockchain Integration in IoT Systems". In: *Journal of Computer Information Systems* 64.2, pp. 265–277. DOI: 10.1080/08874417.2023.2193943. URL: https://doi.org/10.1080/08874417.2023.2193943.

Dinde, Shital and Suresh Shirgave (2024). "Blockchain-IoT Integration: A Systematic Review of Security Challenges and Future Directions". In: *Journal of Electrical Systems* 20.11s.

Fernandez-Carames, Tiago M. and Paula Fraga-Lamas (2024). "A Review on the Use of Blockchain for the Internet of Things". In: *arXiv preprint arXiv:2302.00687*. URL: https://arxiv.org/abs/2302.00687.

Guo, Hao, Wanxin Li, and Mark Nejad (2023). "A Hierarchical and Location-aware Consensus Protocol for IoT–Blockchain Applications". In: *arXiv preprint arXiv:2305.17681.*

Haque, Ehtisham Ul, Adil Shah, Jawaid Iqbal, et al. (2024). "A Scalable Blockchain-Based Frame- work for Efficient IoT Data Management Using Lightweight Consensus". In: *Scientific Reports* 14.1, p. 7841.

Khan, Rafiullah, Sarmad Ullah Khan, Rifaqat Zaheer, &Shahid Khan . (Dec. 2012). "Future Internet: The Internet of Things Architecture, Possie Applications and Key Challenges". In: *2012 10th International Conference on Frontiers of Information Technology*, pp. 257–260. DOI: 10. 1109/ FIT.2012.53.

Kumar, KB Shiva B., Veeresh, & Iwendi, C.. (2024). "To Design and Develop the Hybrid Blockchain-Enabled IoT System for Secured Industry 4.0 Systems". In: *Journal of Smart Internet of Things* 2024.2, pp. 93–105.

Kumar, Nallapaneni Manoj and Pradeep Kumar Mallick (2018). "The Internet of Things: Insights into the building blocks, component interactions, and archi- tecture layers". In: *Procedia Computer Science 132. International Conference on Computational Intelligence and Data Science*, pp. 109–117. ISSN: 1877- 0509. DOI: 10.1016/j.procs.2018.05.170.

Lawad, Suhas and Prashant Nitnaware (2023). "Blockchain Adoption in Real-Time Applications". In: *International Journal of Information Systems and Applied Engineering (IJISAE)* 12, pp. 853–859. Accessed: 2023-12-15. URL: www.ijisae.org/index.php/IJISAE/article/download/6308/ 5114/ 11437.

Lin, Jie et al. (2017). "A Survey on Internet of Things: Architecture, Enabling Technologies, Security and Privacy, and Applications". In: *IEEE Internet of Things Journal* 4.5, pp. 1125–1142. DOI: 10.1109/JIOT.2017.2683200.

Liu, Shuo (2019). "Application of Blockchain in IoT Data Trust and Information Available Technology". In: *2019 12th International Symposium on Computational Intelligence and Design (ISCID)*, pp. 250–254. DOI: 10 . 1109/ ISCID.2019.10130.

Nakamota, Santoshi (2008). *Bitcoin: A Peer-to-Peer Electronic Cash System.*

Ramezan, Gholamreza and Ehsan Meamari (2024). "zk-IoT: Securing the Internet of Things with Zero-Knowledge Proofs on Blockchain Platforms". In: *arXiv preprint arXiv:2302.08322.*

Rangwala, Murtaza and Rajkumar Buyya (2024). "TrustMesh: A Blockchain-Enabled Trusted Distributed Computing Framework for Open Heterogeneous IoT Environments". In: *arXiv preprint arXiv:2411.13039.*

Ray, Partha Pratim (2018). "A Survey on Internet of Things Architectures". In: *Journal of King Saud University—Computer and Information Sciences* 30.3, pp. 291–319. ISSN: 1319-1578. DOI: 10.1016/j.jksuci.2016.10.003.

Chapter 5

Intelligent Blockchain of Things

Enhancing trust and efficiency in IoT ecosystems

Kuldeep Singh Kaswan, Jagjit Singh Dhatterwal, Kiran Malik, and Arvind Panwar

5.1 INTRODUCTION

The IoT is a fast-growing network that has linked devices together and made their interactions automatic across various sectors. However, the development of smart and bigger and complex IoT networks pose challenges concerning trust, security and efficiency. To solve these problems, there is a potential in using the concept of the incorporation of Blockchain technology in IoT systems, namely, the IBoT.

Internet of Things can be described as a network of devices that are connected with each other and exchange data using the internet. These devices may vary from just a sensor to a sophisticated device and form a huge web. Other industries, including healthcare and manufacturing among others, have experienced a major transformation through IoT through the real-time data gathering, monitoring and managing. The expansion of the IoT networks is, however, associated with several challenges, including data security, privacy and system functionality due to the massive expansion of the devices (Bangare and Patil, 2022).

The need for trust and efficiency in IoT ecosystems

For such large-connected environments, proper protection and trust as well as optimal functioning of these connections become crucial with the development of IoT ecosystems. Trust is important because many IoT devices deal with personal information, and as a result, attractive to hackers (Brotsis et al., 2021). These challenges cannot be solved by traditional centralized security models since they are not scalable and have a number of troublesome issues such as having single points of failure. Furthermore, the response time of IoT systems is very important to support real-time analytics of data to make the right decision. There can be issues with delay or inefficiency in communication that can result in large operational disruptions where time-sensitive data is relevant (Kaswan et al., 2023).

DOI: 10.1201/9781003584728-5

Introduction to Blockchain technology

Blockchain is a decentralized, transparent and secure approach to record all the transactions that may take place in the IoT ecosystems; it is suitable for IoT's trust requirement Basically, unlike a conventional database, it is a distributed digital ledger that maintains a continuously growing list of records, called blocks, linked and secured by cryptography and shared by participants across a network (Karaata, 2019). The decentralization of the Blockchain assets minimizes the exposure to cyber attackers and thus enhances the solidity of the network. Moreover, due to its property of being a distributed ledger technology, the data that is recorded on Blockchain cannot be changed or deleted; thus, it offers a high level of data integrity. For these reasons, Blockchain can be effectively applied to IoT to meet the trust and security requirements.

Intelligent Blockchain of Things (IBoT)

The IBoT model integrates Blockchain with IoT to offer a much safer system, which is also self-governing and much more effective. IBoT uses smart contracts, the contracts in which the necessary conditions of the contract are coded (El Masri & Hussain, 2021). These smart contracts can be used within the process of controlling the IoT networks and cut the role of human interference as far as possible. In addition, IBoT utilizes decentralized autonomous organizations (DAOs) for IoT devices management. Thus, IBoT can work more effectively and flexibly. The use of block chain technology also provides ways of increasing the security in IoT as well as improving the functionality of IoT ecosystems.

All these form the pillars from which the potential of IBoT can be understood. The remaining part of the chapter will further discuss the relationship between Blockchain and IoT and how this combination can solve current difficulties and create new possibilities for advancement. The topics, which will be discussed using different fields and presenting the future prospects of IBoT, will give a systematic summary of the effect of IBoT for multiple fields (Ferrag et al., 2018).

5.2 THE INTERSECTION OF BLOCKCHAIN AND IOT

Blockchain has been integrated with the IoT in a way that has totally altered the way connected devices operate in a network. When crossing this technology with IoT, the new opportunities created by the marriage between the decentralized, secure and transparent Blockchain with the IoT's ubiquity and wealth of data generation capabilities can provide the basis for more robust systems. It is important here to understand how these two relate in order to counter the immediacies of IoT while simultaneously using the advantages that are afforded by Blockchain to boost the security, efficiency and independent function (Maroufi et al., 2019).

5.2.1 Synergy between Blockchain and IoT

The areas where Blockchain and IoT intersect are due to their strengths and how they can support each other. IoT devices produce massive data, and this data needs to be stored, processed and shared safely and efficiently. Blockchain offers a distributed ledger that allows establishing data consistency and openness.

Thus, they are suitable for using IoT-produced data. Integrating Blockchain increases the security protocols of IoT networks because Blockchain comes with a feature of tamper-proof ledger system. This synergy also creates a scope for decentralization as Blockchain can allow the IoT smart contracts that can semi-automate the process within the IoT environment and minimizes the need for controlling authorities (Nguyen et al., 2019).

5.2.2 Benefits of integrating Blockchain with IoT

Integrating Blockchain with IoT offers several significant benefits:

- *Enhanced security*: One of the flaws in traditional IoT systems is that they usually have a point of control while using Blockchain means that IoT devices will not be vulnerable to cyber hackers. Every transaction that occurs within the network is encrypted and tied to the previous transactions, making it almost impossible to alter or tamper with.
- *Data integrity and transparency*: Blockchain guarantees that once information is put down, it can't be edited. This immutability is fundamental in preserving the quality of IoT data since it is generally employed in decision-making. Moreover, the mechanism of decentralization by design implies that all the participants in the network can control all the processes and activities, which increases confidence between the participants.
- *Decentralization and autonomy*: Blockchain supports decentralized networks, enabling IoT devices to operate autonomously without reliance on centralized authorities. Smart contracts further enhance this autonomy by automatically executing predefined actions when specific conditions are met, streamlining operations and reducing human intervention (Almarri and Aljughaiman, 2024).

However, there are several challenges when adopting Blockchain technology in IoT systems. One is the problem of scalability in that the more transactions occur on a Blockchain network, the slower the network becomes, and this is expected in IoTs. Blockchain also has a high computational demand that may put additional pressure on IoT devices, especially those that have low potency and power capacity. Moreover, issues related to compatibility of the Blockchain with the prevailing IoT environment and the legal and governance issues of the decentralized structure of the network have emerged. These are some of the concerns that need to be met if Blockchain use in IoT is to be successful.

5.3 ARCHITECTURE OF INTELLIGENT BLOCKCHAIN OF THINGS

IBoT architecture involves the integration of Blockchain technology into the IoT, where the requirements of a secure, decentralized and efficient system are met. It is based on several main elements that provide improved functionality to the IoT networks to make them smarter and more autonomous. To grasp the IBoT taxonomy and to establish proper solutions that can fix the issues of IoT, there is a need to understand the key system components (Tian & Tian, 2024).

5.3.1 Core components of IBoT

The major components of IBoT architecture include IoT devices, Blockchain nodes, smart contract and consensus mechanism. IoT devices are the data producers in the network, and they collect information from the surroundings that are logged into the Blockchain. The nodes of the Blockchain include a record of procedures, which are located and distributed in other nodes of the network, to guarantee visibility and non-amendment. Smart contracts are self-contained programs that run on the basis of scripts that trigger overt and specific actions depending on given triggers or inputs. The consensus mechanism is vital in the process of validating transactions and the consistency of the Blockchain at that nodes in a system are interchangeable and all have the same perception of the ledger (Wang et al., 2013).

5.3.2 Smart contracts and IoT devices

Smart contracts are critical to the IBoT success since they help simplify the IoT processes within the ecosystem. These contracts are automated and programmed with predetermined guidelines that generate actions on predefined circumstances such as signing of transactions or managing IoT gadgets. Smart contracts when connected with IoT devices enable real-time decision-making and increase the efficiency of operations. For instance, in a smart home, a smart contract could manage the temperature and regulate the use of home heating systems with the help of temperature sensors without involving people.

5.3.3 Consensus mechanisms in IBoT

Blockchain consensus mechanisms form the core of the IBoT architecture safety and operation. They are responsible for making sure that all the nodes of the network have the same opinion on problems and validity of the transactions for the sake of avoiding any distortion of the Blockchain. The consensus that can be used in IBoT systems is as follows: Proof of Work (PoW), Proof of Stake (PoS) and Practical Byzantine Fault Tolerance (PBFT). All these mechanisms have their pros and cons: while PoW is highly secure, it can be very costly, both in terms of time and resources; PoS, on the other hand, is very efficient in terms of time, but much depends on the trust that can be placed in its stakeholders. The type of consensus mechanism to be employed for IBoT application depends on the requirements of the particular application (Dedeoglu et al., 2019).

5.3.4 Decentralized autonomous organizations (DAOs) for IoT

Decentralized autonomous organizations (DAOs) as a form of managing the IoT networks is a relatively new approach in the context of IBoT. DAOs work without any central authorities and control the functioning and actions using certain rules made within smart contracts. DAOs can also seamlessly manage several devices in the context of IoT in that they do not require the supervision of human beings much of the time. For example, a DAO may decide the routes and the charging times of a fleet of self-driving cars upon certain goals, or objectives, using data input and output from the cars themselves, without any human interference.

5.3.5 Scalability solutions for IBoT

One of the main deployment challenges relates to scalability because integrating Blockchain with IoT results in elevated volumes of data and transaction rates (Kaswan et al., 2024a). To fix this, numerous scalability mechanisms have been devised, including but not limited to sharding, off-chain transactions and layer 2 solutions. Sharding refers to the process by which the Blockchain is split into smaller parts, making it possible to process transactions concurrently. Off-chain transactions unload some transactions and keep them separate from the main Blockchain and layer-2 protocols like lightning network facilitate the rapid and efficient processing of the transactions. To apply these solutions, it is crucial in order to achieve the demands to enforce IBoT systems that will need to the growing IoT applications in huge scale (Yang et al., 2022).

5.4 ENHANCING TRUST IN IOT ECOSYSTEMS

Security is one of the most crucial obstacles in the IoT due to the connectivity of different objects and the confidentiality of the information processed. With the continual implementation of the IoT devices, the challenge on how to secure and create trust among the devices rises. Blockchain technology presents a viable solution whereby it brings about decentralized, immutable and transparent method of authentication and authorization in these ecosystems in order to increase trust in data sharing. In this particular section, it is proposed how Blockchain can be utilized to solve trust-related issues in an IoT setting with particular focus on different sectors.

5.4.1 Blockchain-based authentication and authorization

Secure authentication and authorization are crucial in any system and more so in IoT environments where devices are self-actuating. Conventional approaches to these processes are centralized and have become unfavorable targets for hacking and are full of "single node failure" points. The solutions on the basis of the Blockchain allow increasing the levels of the IoT devices' authentication and authorization, which improve the network security and reliability. Through the nature of attributes and consensus, the devices apply for identity and permission for the interaction with other devices or services without involving authority. This decentralized structure does not only lessen the probability of intrusion by unwanted parties but also leaves no one party capable of bringing down the whole system (Rathee et al., 2023).

5.4.2 Trustless data sharing and transactions

Another benefit of the Blockchain technology is that it enables the sharing of information and performing transactions where other parties do not necessarily need to be trusted. As it will be described in the sections below, while a central entity manages the IoT data in a more conventional IoT architecture, there are problems concerning data accuracy, ownership, confidentiality and security. Traditional centralized systems are not suitable for personalized data sharing due to various constraints such as security, privacy and the need for middleware, all these concerns are effectively addressed in Blockchain by virtue of peer-to-peer (P2P) sharing of data. Every transaction or exchange of data is noted in the Blockchain to allow for reviews when there is a need, and the results of such reviews can be as well seen by all parties. This trustless model assures that data is not forged and that all transactions are executed in compliance to smart contracts enhancing trust in IoT environment.

5.4.3 Mitigating IoT vulnerabilities through Blockchain

IoT devices become appealing for hackers since those devices possess fewer computational capacities as well as less developed security procedures. IoT edges can, however, be made less vulnerable to these vulnerabilities by the use of Blockchain as the underlying environment for IoT communications. A feature of the Blockchain is encryption through the use of cryptographic techniques that enable devices to exchange encrypted information that can only be decrypted by the authorized entities. Furthermore, since it contains no central point of control, it is almost challenging to destroy the Blockchain by an attacker. This robust security framework is helpful in essential use cases for IoT, which involve catastrophic risk factors, for instance, healthcare solutions and smart city integration (Liu et al., 2020).

5.4.4 Case studies

Trust enhancement in various IoT sectors can happen as in the following instances.

- *Smart cities*: In smart city projects where different IoT devices control and oversee the city's infrastructure, Blockchain can help provide structural reliability or facts related to transport, utilities and security. For example, through the application of block chain some of the objectives that can be achieved include traffic data from vehicles can be guaranteed to be genuine and cannot be forged enabling efficient traffic management.
- *Healthcare*: Blockchain also has application in the protection of the patients' information to allow only a few people access to such data in the healthcare sector. This helps to restrict fraud and maintain the privacy and security of medical information, which is basic for the trust in healthcare.
- *Supply chain management*: In supply chain management, Blockchain increases the level of trust since every owner of the supply chain has a record of the supply chain. It can as well curb fraud and also prove the credibility of products, and, in addition, facilitate supply chain because all the interest parties will be in a position to access equal real-time information.

5.5 IMPROVING EFFICIENCY IN IOT THROUGH BLOCKCHAIN

Scalability is another important criterion because IoT uses a multitude of interconnected devices and the latter must work as efficiently as possible arranging and transmitting information. Introducing Blockchain into the

IoT technology creates a new concept called the Intelligent Blockchain of Things (IBoT), which provides a better boost to this efficiency by minimizing manual interferences, control overheads and consumption of resources. In this section, ways through which Blockchain can enhance IoT operations, enhance efficiency, minimize delays and support real-time data processing are discussed including relevant real-life examples (Kaswan et al., 2024b).

5.5.1 Streamlining IoT operations with smart contracts

Smart contracts are contracts that contain rules and according to which they are performed automatically. They can be used in the IoT ecosystem to decentralize various operations, including the processes of device identification, as well as payments. Smart contracts thus eliminate the need for third parties and middlemen, precisely because they allow machines to communicate directly with each other, operations are simplified and made more efficient. For instance, in the smart home context, a smart contract might automatically fine-tune the usage of energy because of data from IoT sensors, well without involve an actual person.

5.5.2 Energy efficiency in IBoT systems

Energy is another important factor in the IoT because of the vast number of devices that are connected to the network and, in most cases, these devices need power to operate. When applied to IoT, the use of Blockchain technology plays an important role in rationing and, therefore, improving energy efficiency since it does not require an ongoing process of data exchange. IBoT systems can allow energy usage by priority where only important operation is powered when important, while the less important one is powered when the energy is cheap. Also, the concept of smart contracts together with the distributed structure of Blockchain stimulates further energy saving because computing is distributed and does not rely on centralized high-energy servers (Maeng et al., 2022).

5.5.3 Reducing latency and improving speed

Another issue common to IoT systems is latency, understood as the time it takes for the effect of an action to be observed, which is highly important in terms of time sensitivity in most IoT applications. Blockchain can contribute to decreasing latency by enabling more devices to perform processing on their own, eliminating the need to go through a headquarters server. Hence, the direct interaction is not only faster in terms of processing the data but also provide faster and efficient free-running IoT systems for real-time conscious applications like self-driving cars and industries.

5.5.4 Real-time data processing and analytics

Real-time data processing capability holds a key advantage in a number of IoT application scenarios, especially those concerning infrastructure systems. Blockchain technology helps to process data in real time as the data is quickly validated and then will be safely stored, while IoT devices do not waste time to act accordingly. In an IoT system, smart contracts are capable of taking actions that are commensurate with the data analytics and real-time results; an instance is when a machine malfunctions and the smart contract halts the machine so that it cannot further exacerbate the problem and at the same time increase efficiency (Shammar et al., 2022).

5.5.5 Case studies

Efficiency improvements in IoT ecosystems can be accomplished through IBoT.

- *Smart manufacturing*: In smart manufacturing, blockchain IoT helps through the supply chain management as the material is delivered at right time so that there is no high stock. Smart contracts can also manage production to devise the most efficient production regimes, using real-time data from the sensors in IoT.
- *Autonomous vehicles*: While highly automating wheel chauffeurs are still being developed, latency should be minimized and latest information should stay processed. In the Blockchain setup, vehicles have direct interaction with other vehicles or the surrounding infrastructures and as this speeds up the decision-making process, the possibility of a crash due to such delays is eliminated.
- *Environmental monitoring*: In monitoring the environment, Blockchain and IoT will come in handy to make sure that information from the sensors is processed and acted upon instantly. For instance, in the smart irrigation system, Blockchain IoT can automatically control irrigation systems, based on the real-time data of soil moisture and weather condition and thus minimizing fouling and wastage.

5.6 APPLICATIONS OF INTELLIGENT BLOCKCHAIN OF THINGS

The application of blockchain to IoT is revolutionizing several sectors through the provision of reliability, security and real-time solutions. This section discusses specific cases of the use of IBoT in principal sectors where this technology is already changing the nature of industries, such as homes, transport and environmental tracking.

5.6.1 Industry 4.0 and smart manufacturing

In the context of the Fourth Industrial Revolution (Industry 4.0), smart manufacturing is a continuation of the use of IoT, AI and robotics in developing better and flexible solutions in production. The integration of Blockchain into this ecosystem presents a further improvement of the manufacturing process by creating for server delivery of data records in a secure and, more importantly, impenetrable manner. For instance, Blockchain can be applied for supply chain management where every stage, starting from the acquisition of raw materials through to delivery, is documented and validated. Smart contracts remove intermediaries like human beings in executing tasks, such as order execution and payment, among others, and will do them faster. This in turn results to improved productivity, less expenditure, as well as improved confidence from the parties involved (Fu et al., 2019).

5.6.2 Smart homes and connected devices

The use of IoT in homes gives people the ability to control the lighting, heating systems and even the security systems of homes. The integration of the smart homes' systems with the Blockchain technology ensures that all the data being exchanged within the network is safe from any form of manipulation. Blockchain, therefore, brings better security to smart home networks through decentralization of data and allowing secure peer-to-peer communications within devices. For example, based on real-time data, smart contracts can control the use of energy and, in the process, save costs. Further, the concept of Blockchain technology can also protect the privacy of users since the data which is shared by the users will be open only to specific individuals, thereby increasing the level of trust in smart IoT devices.

5.6.3 Autonomous vehicles and intelligent transportation

The automotive industry has transitioned toward using autonomy and intelligent transportation system in cars. The detailed solution of the interactions of these systems is provided by the application of the principles of the Blockchain. In approved vendor list, the data in question might include attributes concerning the identity of the vehicle, its ownership, as well as its history of service and repair; in this context, Blockchain functionality helps to ensure that the information so crucial is both accurate and tamper-proof. However, Blockchain technology can also enhance vehicle-to-vehicle and vehicle-to-infrastructure, interaction providing real-time data for decision-making that is important for safety and efficiency of transportation. For instance, in intelligent transportation systems, Blockchain can better organize traffic, which will decrease the density of cars on the roads and make traffic more efficient (Kaur & Ali, 2021).

5.6.4 Environmental monitoring and sustainability

The application of IoT devices in environmental monitoring and sustainability includes collection of data on environmental factors, including air quality, water levels and temperature. Using Blockchain with IoT in that way guarantees the reliability, confidentiality and genesis of the gathered information. Blockchain can also help maintain a permanent record of environmental information for assessment of compliance with environmental laws and policies as well as support green agenda. For instance, smart IoT-based Blockchain can control energy consumption in intelligent electrical networks and optimize the usage of renewable energy. Also, it can be used to establish the mostly unheard-of decentralized carbon credit trading platforms, which will help to make organizations and individuals cut down their emission levels.

5.7 CHALLENGES AND FUTURE DIRECTIONS

Several issues require an understanding as the combination of Blockchain technology with the IoT becomes more established to provide IBoT solutions. This section provides information on the technical issues that arise when deploying IBoT, legal and regulatory issues that surround IBoT technology, ethical issues involved and future work and prospects in the development of the IBoT.

5.7.1 Technical challenges and solutions

Blockchain and IoT integration faces multiple technical issues, which are mainly confined to scalability, connectivity and energy issues. The current Blockchain networks, especially the public ones, are struggling with the scalability issue because consensus models such as PoW are time-consuming and resource-consuming. Also, IoT devices are inherently dissimilar in terms of communication and data exchange and this brings difficulties when trying to incorporate them into one block chain-based system. Another factor is energy requirement; IoT devices are usually low-power devices and the processing required by Blockchain is burdensome for the usually low-power resources that are in IoT devices. For these issues, possible solutions include off-chain computation and lightweight consensus approach such as the Proof of Stake (PoS) and Edge computing to improve the scalability of IBoT systems.

5.7.2 Regulatory and legal considerations

The implementation of IBoT is therefore not without its consequences as it also introduces regulatory and legal concerns such as data privacy, security as well as compliance with existing laws. In the Blockchain system, many parties are involved, and that makes it challenging to enforce various regulations since there are no authorities controlling them. Also, IoT shares geographical boundaries in many of its applications, which gives rise to the issue of conflicting jurisdictions as the laws on data protection are not always uniform across the globe. It is, therefore, imperative to discuss and lay down the best practices in sourcing and implementation of Blockchain solutions so as to enhance compliance with regulations while at the same time retaining the decentralized advantages of the technology. It is, therefore, important that governments, industries and regulatory authorities work together in order to establish the right environment for innovations that will at the same time enhance the protection of users.

5.7.3 Ethical implications of IBoT

Privacy, surveillance and misuse are critical ethical issues that need to be addressed about IBoT. With many devices using the IoT generating tons of data, hacking and breaches lead to unauthorized access of the data causing problems to the users' privacy. Further, the decentralization of this data through the application of Blockchain software stirs up issues of ethics in data ownership and management. Thirdly, there is a probability of the introduction of IBoT for surveillance purposes, which may cause a threat of loss of independence and freedom of individuals. Mitigating all these ethical issues will entail designing and implementing sound ethical principles and norms that will safeguard user rights and ensure that the Fourth Industrial Revolution's effects of IBoT percolate to all users in society.

5.7.4 Future trends and research directions

In the future, IBoT will also be influenced by the development of technology, the changing legal requirements and the increasing demands for safe and effective IoT systems. The future research will have a tendency toward designing new Blockchain protocols with higher efficiency in terms of energy consumption, IoT devices integration and design of new decentralized applications based on the synergies of both Blockchain and IoT. Also, there will be a rise in research on the social, ethical and economic aspects that arise in the wake of popular use of IBoT. That is why the further development of these trends, as well as the application of artificial intelligence for the management of Blockchain and IoT, the further development of quantum computing in combination with IBoT, will be among the most important trends in the future of this field.

5.8 CONCLUSION

This chapter discusses how IoT can be integrated with Blockchain to address key issues such as security and trust, transparency and efficiency. IBoT aims at overcoming such challenges in IoT networks as data manipulation, unauthorized access and ineffective functioning due to the Blockchain nature of the project. This chapter deals with how more reliable and scalable IoT ecosystems are formed with smart contracts, decentralized autonomous organizations (DAOs) and consensus mechanisms. It also underlines that these innovations are essential in sectors like smart cities, healthcare or supply chain management in which trust and efficiency are of high value. However, the present chapter has also discussed a number of issues that may arise during the implementation of IBoT. Challenges related to technique include the scale, energy consumption and integration and those that are regulatory and ethical in nature are covered in detail. The chapter reviews how some of the challenges could be managed or addressed and stresses that more work is needed. Finally, based on this perspective, the chapter identifies IBoT as having a great potential in developing the IoT environments. In the near future, as Blockchain and IoT technologies develop further, IBoT will serve as a significant enabler of the increased security and optimization of connected devices. Finally, in this chapter, it is stressed how IBoT can revolutionize IoT networks and showing the directions that signpost current technological developments in various fields.

REFERENCES

Almarri, S., & Aljughaiman, A. (2024, November). Blockchain technology for IoT security and trust: A comprehensive SLR. *Sustainability*, *16*(23), 10177. [Online]. Available: https://doi.org/10.3390/su162310177

Bangare, P. S., & Patil, K. P. (2022, April). Security issues and challenges in Internet of Things (IoT) system. in *Proceedings of the 2nd 2022 International Conference on Advance Computing and Innovative Technologies in Engineering (ICACITE)*, Greater Noida, India, pp. 6–11. https://doi.org/10.1109/ICACITE53722.2022.9823709

Brotsis, S., Limniotis, K., Bendiab, G., Kolokotronis, N., & Shiaeles, S. (2021). On the suitability of blockchain platforms for IoT applications: Architectures, security, privacy, and performance. *arXiv*. https://doi.org/10.48550/arXiv.2109.02352 reddit.com+2arxiv.org+2en.wikipedia.org+2ijrcait.com

Dedeoglu, V., Jurdak, R., Putra, G. D., Dorri, A., & Kanhere, S. S. (2019). A trust architecture for blockchain in IoT. *arXiv*. https://doi.org/10.48550/arXiv.1906.11461 dl.acm.orgarxiv.org+1dl.acm.org+1

El-Masri, M., & Hussain, E. M. A. (2021). Blockchain as a mean to secure Internet of Things ecosystems—A systematic literature review. *Journal of Enterprise Information Management, 34*(5), 1371–1405. https://doi.org/10.1108/JEIM-12-2020-0533 emerald.com

Ferrag, M. A., Derdour, M., Mukherjee, M., Derhab, A., Maglaras, L., & Janicke, H. (2018). Blockchain technologies for the Internet of Things: Research issues and challenges. *arXiv*. https://doi.org/10.48550/arXiv.1806.09099 reddit.com+5arxiv.org+5arxiv.org+5

Fu, S., Zhao, L., Ling, X., & Zhang, H. (2019, May). Maximizing the system energy efficiency in the blockchain based Internet of Things. in *Proceedings for the 2019 IEEE International Conference on Communications (IEEE ICC 2019)* Shanghai, China, pp. 1–6. https://doi.org/10.1109/ICC.2019.8761539

Karaata, M. H. (2019, June). An open modal stabilizing-wave algorithm for arbitrary networks. *Journal of Parallel and Distributed Computing, 128*, 15–29. https://doi.org/10.1016/j.jpdc.2019.01.014

Kaswan, K. S., Dhanda, S. K., Dhatterwal, J. S., & Balusamy, B. (2024b, January). Blockchain technology in HR processes. in *The Role of HR in the Transforming Workplace: Challenges, Technology, and Future Directions* (1st ed., Editor Anamika Pandey, Simon Grima, Suruchi Pandey, Balamurugan Balusamy). Publisher Productivity Press, pp. 104–116. https://doi.org/10.4324/9781003372622-8. ISBN: 978-100384604-8, 978-103244532-8.

Kaswan, K. S., & Dhatterwal, J. S. (2024a, January). Blockchain for efficient and improved strategy in power stations. in *Smart Mobility and Intelligent Transportation Systems for Commercial and Hazardous Vehicles* (Editor Naga Swetha Pasupuleti, T. Poongodi, and Balamurugan Balusamy). Apple Academic Press, pp. 99–123. ISBN: 978-100383695-7, 978-177491538-7.

Kaswan, K. S., Sidhu, G. S., Dhatterwal, J. S., & Kumar, N. (2023, October). The role played by blockchain computing premium payments and creating new markets. in *The Application of Emerging Technology and Blockchain in the Insurance Industry* (Editor Adarsh Garg, Fadi Al-Turjman, and Dr. Shahnawaz Khan). IGI Global Scientific, pp. 163–178. ISBN: 978-877004005-1, 978-877004006-8.

Kaur, R., & Ali, A. (2021). A novel blockchain model for securing IoT based data transmission. *International Journal of Grid and Distributed Computing, 14*(1), 1045–1055.

Liu, H., Han, D., & Li, D. (2020). Fabric-IoT: A blockchain-based access control system in IoT. *IEEE Access, 8*, 18207–18218. https://doi.org/10.1109/ACCESS.2020.2968492 journals.orclever.com

Maeng, J., Heo, Y., & Joe, I. (2022). Hyperledger Fabric-based lightweight group management (H-LGM) for IoT devices. *IEEE Access, 10*, 56401–56409. https://doi.org/10.1109/ACCESS.2022.3177270 journals.orclever.com

Maroufi, M., Abdolee, R., & Mozaffari Tazekand, B. (2019). On the convergence of blockchain and Internet of Things (IoT) technologies. *arXiv* .https://doi.org/10.48550/arXiv.1904.01936 arxiv.org

Nguyen, D. C., Pathirana, P. N., Ding, M., & Seneviratne, A. (2019). Integration of blockchain and cloud of things: Architecture, applications and challenges. *arXiv*. https://doi.org/10.48550/arXiv.1908.09058 arxiv.org

Rathee, G., Ahmad, F., Jaglan, N., & Konstantinou, C. (2023). A secure and trusted mechanism for industrial IoT network using blockchain. *IEEE Transactions on Industrial Informatics, 19*(2), 1894–1902. https://doi.org/10.1109/TII.2022.3182121 journals.orclever.com

Shammar, E. A., Zahary, A. T., & Al-Shargabi, A. A. (2022). An attribute-based access control model for Internet of Things using Hyperledger Fabric blockchain. *Wireless Communications and Mobile Computing, 2022*, 6926408. https://doi.org/10.1155/2022/6926408 journals.orclever.com

Tian, H., & Tian, J. (2024). A blockchain-based access control scheme for reputation value attributes of the Internet of Things. *Computers, Materials & Continua, 78*(1), 1297–1310. https://doi.org/10.32604/cmc.2024.047058 techscience.com

Wang, J., Bin, S., Yu, Y., & Niu, X. (2013). Distributed trust management mechanism for the Internet of Things. *Applied Mechanics and Materials*, 347–350.

Yang, X., Yang, X., Yi, X., Khalil, I., Zhou, X., & He, D. (2022, March). Blockchain-based secure and lightweight authentication for Internet of Things. *IEEE Internet of Things Journal*, *9*(5), 3321–3332. https://doi.org/10.1109/JIOT.2021.3098007

Chapter 6

Advanced privacy preservation in IoT using Blockchain and multi-objective optimization

Poonam Samir Jadhav and Gautam M. Borkar

6.1 INTRODUCTION

The Internet of Things (IoT) plays a significant role in connecting sophisticated devices on the Internet, which accelerates data transfer between distinct places with the utilization of diverse protocols. In recent years, the necessity of IoT-based techniques has increased in various real-time environment domains, such as financial services, healthcare, traffic monitoring, manufacturing, energy transfer, and weather monitoring [1,2]. From this perspective, a recent survey of International Business Machines (IBM) states that the utilization of IoT-based devices in 2027 will reach more than 1.4 trillion, because of its compactness and minimal power consumption [3]. Furthermore, IoT technologies are more effective in sensing every individual in the network with the existence of resource constraints, standard protocol, and centralized repository. How- ever, the devices are significantly vulnerable to smart environments, security issues, and privacy breach problems [2,4,5]. To address these security issues, Blockchain is the most prominent factor for ensuring data integrity in IoT applications. In this context, the Blockchain contains several blocks such as block header, transaction data, previous block hash, and timestamp for accessing the transaction copy of every node. Generally, Blockchain technology provides various advances as traceability, tamper-resistant, consistency, transparency, pseudonymity, and improved efficiency [2,6,7]. However, the major problems are scalability, anonymity, and compromising privacy [8,9]. Therefore, an improved security scheme is implemented to prevent the loss of privacy problem during data transmission [7]. According to this, privacy preservation provides crucial impacts on data collection, processing, and sharing the personal information through data nodes [10]. Therefore, the standard technical methods that ensure user privacy data employ encryption, anonymization, differential privacy, masking, secure enclaves, and pseudonymization. In spite of these techniques, anonymization models achieve high-security environments with simple algorithmic and data analysis functions [11]. Numerous anonymization techniques were obtained to safeguard privacy and sensitive

 DOI: 10.1201/9781003584728-6

information [12], such as k-anonymity, partial suppression, L-diversity, and generalization [13]. Conceptually, k-anonymity enhances data security with the combination of suppression and generalization that enhances the transaction data without disclosing personal sensitive information [14]. However, these conventional methods suffered from certain disadvantages as inconsistent identification, data threats, complexity issues, and disclosure risks specifically [12]. The Canine optimization-based k-anonymization-enabled privacy preservation scheme (COKAPS) scheme is efficiently used to evaluate the privacy leakage problem in IoT-based network systems. Conceptually, the COKAPS scheme achieved better privacy and security during data sharing, and the performance effectiveness is validated under various privacy measures as utility, privacy, and genuine index. In the research, the data provider and end-user garnered the specific data with a high range of privacy measures. The major contributions of the research are portrayed below,

- Canine optimization (CAeO)
  To improve the effectiveness and ability of the privacy preservation tasks, the CAeO algorithm is introduced, which provides an efficient optimal value by providing a better convergence rate and thereby enables to reduction of optimization problems.
- COKAPS
  The k-anonymization technique is induced to prevent the privacy disclosure risk, which effectually garnered private data from data providers and preserves the data in a cloud platform without causing any security threat. Thus, the k-anonymization process is carried out by generalization and suppression methods.

6.1.1 Motivation

The increasing integration of IoT devices into critical sectors such as healthcare, finance, and urban infrastructure has heightened the risk of security breaches, data leakage, and scalability bottlenecks. Although Blockchain technology and existing anonymization techniques offer partial solutions, they often suffer from high computational complexity, limited scalability, and insufficient privacy guarantees in dynamic environments. These limitations highlight the need for a more effective, scalable, and privacy-preserving framework for Blockchain-enabled IoT systems.

6.1.2 Contribution

This chapter makes the following key contributions:

- Proposes a novel privacy-preserving framework named COKAPS (canine optimization-based K-anonymization privacy scheme) designed specifically for Blockchain-enabled IoT systems

- Introduces a canine-inspired multi-objective optimization algorithm to enhance the quality of anonymization by optimizing privacy, authenticity, and utility factors
- Validates the proposed approach through comprehensive simulation studies using real-world datasets, demonstrating significant improvements over traditional methods in terms of privacy assurance, system utility, and scalability
- Provides a scalable, efficient, and adaptable solution for ensuring secure and privacy-aware data management in diverse IoT application scenarios.

The remainder of the chapter is arranged as follows: Section 6.2 describes the summary of recent literature articles on privacy preservation in anonymization techniques along with problem formulation. Section 6.3 presents the system model of the IoT-based Blockchain technology models. Section 6.4 elaborates the proposed methodology with a corresponding anonymization mechanism. Section 6.5 discusses the results obtained from the proposed model and proves the feasibility of advances in real-time environment applications. Finally, Section 6.6 concludes the chapter describing the future outlook.

6.2 LITERATURE REVIEW

The section provides a detailed summary of various traditional methods in the field of anonymity-based privacy preservation techniques that are discussed effectively.

Arun Kumar Sangaiah et al. [15] established a k-anonymity-based privacy preservation model for validating the security and privacy threats problems. In the method, the private information of users was secured by the entropy evaluation. But the unauthorized access in the data center generated an issue as a data mining and re-identification problem.

Anh-Tu Hoang et al. [16] developed a personalized cluster-based anonymization algorithm with the integration of a knowledge graph (PCKGA) to preserve the private information of data providers. Furthermore, the model effectively generates qualified anonymized KG with the utilization of k-attribute degree (p-k-ad). However, the PCKGA model suffered from imbalanced data issues, privacy disclosure risks, and over-anonymization issues.

Abdul Majeed et al. [12] introduced an ML-based privacy preservation model to evaluate the vulnerability attacks that were induced in the network. The method efficiently improved the utility equilibrium and privacy equilibrium for anonymized data, whereas the latency and throughput issues were obtained because of the inaccurate and imbalanced dataset. Therefore, a new model was introduced in the future to overcome the above-mentioned limitations effectively. Jun Ha Lee and Su Jeong You [14] implemented DL-based privacy preservation techniques under the utilization of computer vision tasks, which evaluate the optimal value of anonymization efficiently. In the method, the personal information of the users was secured by anonymization pattern evaluation. Based on these executions, the model attained effective recognition accuracy with better privacy protection. Furthermore, the major impact obtained in the model was performance degradation, data complexity issues, and generalizability problems.

Adla Padma et al. [2] employed a SecPrivPreserve framework to ensure the integrity and security of IoT networks. In the method, the sensitive data were protected by various preservation techniques such as encryption, one-time pass- words, and hashing. Based on these techniques, the model effectively improved integrity, confidentiality, and privacy. However, the model faced performance degradation, scalability issues, and potential problems.

Noe Elisa et al. [17] established a decentralized Blockchain framework for evaluating the efficacy of privacy data preservation. Additionally, the Blockchain integrity of the model was evaluated by the artificial immune system that significantly analyzed the insider and external threats. During the evaluation, the operational time was increased and caused imbalance and scalability challenges.

Yedida Venkata Rama Subramanya Viswanadham and Kayalvizhi Jayavel [9] introduced an ethereum Blockchain framework with the integration of the adaptive border Collie rain optimization algorithm (ABC-ROA) for effective validation of privacy preservation. The model reduced the complexity problem and preserved the transmitted data efficiently. Further, the data sanitization and restoration were performed by an optimization algorithm, which generates an optimal key for multivalidated constraints. However, the hiding failure and false rule generation of the model were increased during evaluation and faced high error rate problems.

Shan Jiang et al. [18] developed a bloom filter-based multi-keyword protocol for evaluating network privacy preservation. In the method, the multi-keyword protocol defines the deletion and addition of protocol in dynamic data record updation. Because of these updates, the developed model reduced the query time and computational cost effectively. While evaluating large-scale data, the model's performance was degraded, and varied encryption types to preserve the private data in future directions.

J. Andrew et al. [19] implemented a clustering-based k-anonymization model for securing the network from identity disclosure attacks. Meanwhile, the model effectively detects the privacy attacks caused by data providers and data collectors. Due to this, the model effectually tackles the external and internal disclosure risks with precise data protection and minimal complexity issues. But the model suffered from other disclosure threats as similarity attacks, attribute disclosure, and membership disclosure, which reduced the data utility gradually in the network.

Wei Zhou et al. [20] employed a decentralized Blockchain framework with the integration of an adaptive differential privacy algorithm (MDLDP) to preserve private data effectively. The method significantly reduced the potential issues and strengthened the transactional security gradually. Furthermore, the developed model boosts data availability with enhanced privacy protection. However, the model faced data validity problems, latency issues, and lack of data information specifically.

6.2.1 Challenges

The challenges encountered in the conventional methods are detailed below:

- The k-anonymity-based privacy preservation model faced unauthorized personnel access in the network that caused data mining problems and privacy disclosure risks [15].
- When validating large size of data information, the bloom filter-based multi-keyword protocol degrades the performance effectiveness with maximal computational resource constraints [18].
- The security measure parameters of utility and privacy were reduced in the clustering-based k-anonymization process. By considering these reductions, the network suffered from various internal disclosure risks that caused a high error rate effectually [19].
- Under the evaluation of MDLDP, the model faced certain disadvantages such as latency problems, imbalance issues, data validity problems, and complexity issues [20].
- While evaluating the ABC-ROC model, the network suffered from increased hiding failure, and false alarm rate generalization, which caused an impact on performance effectiveness [9].

6.2.2 System model of IoT

Figure 6.1 illustrates the system model of the IoT structure that creates a detailed representation of various system technologies to achieve secure end-to-end communication with modularity, scalability, behaviors, configurations, and interactions of IoT deployments in diverse environments. In this, the four major layers of IoT technology are the sensor layer, networking layer, cloud servers, and application layer [21], which are briefly explained in the following sections.

- *Sensor layer*: This sensor layer contains a large amount of smart objects with the interconnection of various sensor types such as motion, color, flame, and camera. Conceptually, the sensor computes the physical property and converts it into signal property. Meanwhile, the sensor objects are effectively integrated with sensing information

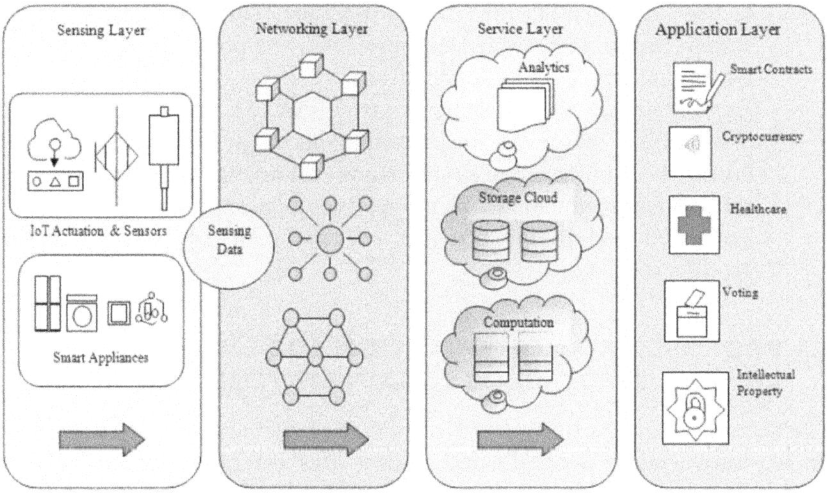

Figure 6.1 System model of IoT structure.

to attain accurate decisions of monitoring and recording in the evaluation of capacity measurement such as air quality, pressure, temperature, humidity, electricity, speed, flow, and movement [22]. Therefore, the attained sensors are grouped together and form a unique behavior, which is then forwarded into the sensor gateway to generate effective productivity with improved battery life.

- *Networking layer*: The data transition from source to destination nodes is performed effectively with wired or wireless network transport medium. The network generates significant communication needs such as bandwidth, latency, and security. Additionally, in the network layer, Blockchain technology also is present, which contains several blocks that are interconnected to record the transaction of decentralized and secure data along with the information of previous data transition. Furthermore, the high range of transition requires star, hybrid, tree, and mesh network models to access the protocol in heterogeneous configuration and provides effective benefits in various applications such as transactional services, context-aware applications, and so on. However, the network models have some drawbacks such as energy consumption, malicious attacks, and data throughput.
- *Service layer*: The management service layer presents effective information processing by security controls, analytics, device management, and process modeling. In this, the privacy disclosure threats of data sources are reduced by data filtering techniques such as data

synchronization, data anonymization, and data integration, which secures the essential information precisely. Thus, the security of the system restricts unauthorized personnel, the possibility of risk, and system hacking problems effectively.

– *Application layer*: The massive growth of IoT technology provides effective advantages in various domains from healthcare to smart homes. Some of them are retail supply chain, agriculture, user interaction, environment and energy, building, culture, and tourism. Thus, IoT technology provides effective reliability with improved product efficiency.

6.3 PROPOSED K-ANONYMIZATION FOR ENSURING PRIVACY IN IOT -BLOCKCHAIN SYSTEMS

The section describes the privacy preservation technique developed based on k-anonymization with the integration of a canine optimization algorithm. Generally, in computing systems, multiple numbers of data owners send their data to cloud servers to secure the data information and require easy retrieval for future purposes. Meanwhile, the confidential data that is secured in the cloud platform is effectively captured by the end users. However, the major challenges encountered in data transmission among information providers and end users are latency, energy consumption, and compromising privacy. To resolve these drawbacks, a k-anonymization-based privacy preservation technique enabled by the canine optimization algorithm (COKAPS) is proposed.

Privacy preservation is an essential concern in cloud network that captures, processes, and shares confidential information securely. In this network, vulnerability attacks and unauthorized personnel are significantly restricted by various privacy preservation techniques. In the research, the sensitive information of the data providers is highly secured by k-anonymization-based privacy preservation tasks. Moreover, the k-anonymization effectively solved the privacy compromise threats with minimal time consumption and preserved the complex data information. Because of these advances, k-anonymization-based privacy preservation is introduced in this research, which provided robust and generalized results precisely. Thus, the diagrammatic representation of the COKAPS scheme is illustrated in Figure 6.2.

6.3.1 Data representations

The multiple data owners transfer their original information to an information database, which contains diverse records and attribute functions to perform the storage processes. Let us consider that the information database contains a distinct number of records, which are effectively evaluated

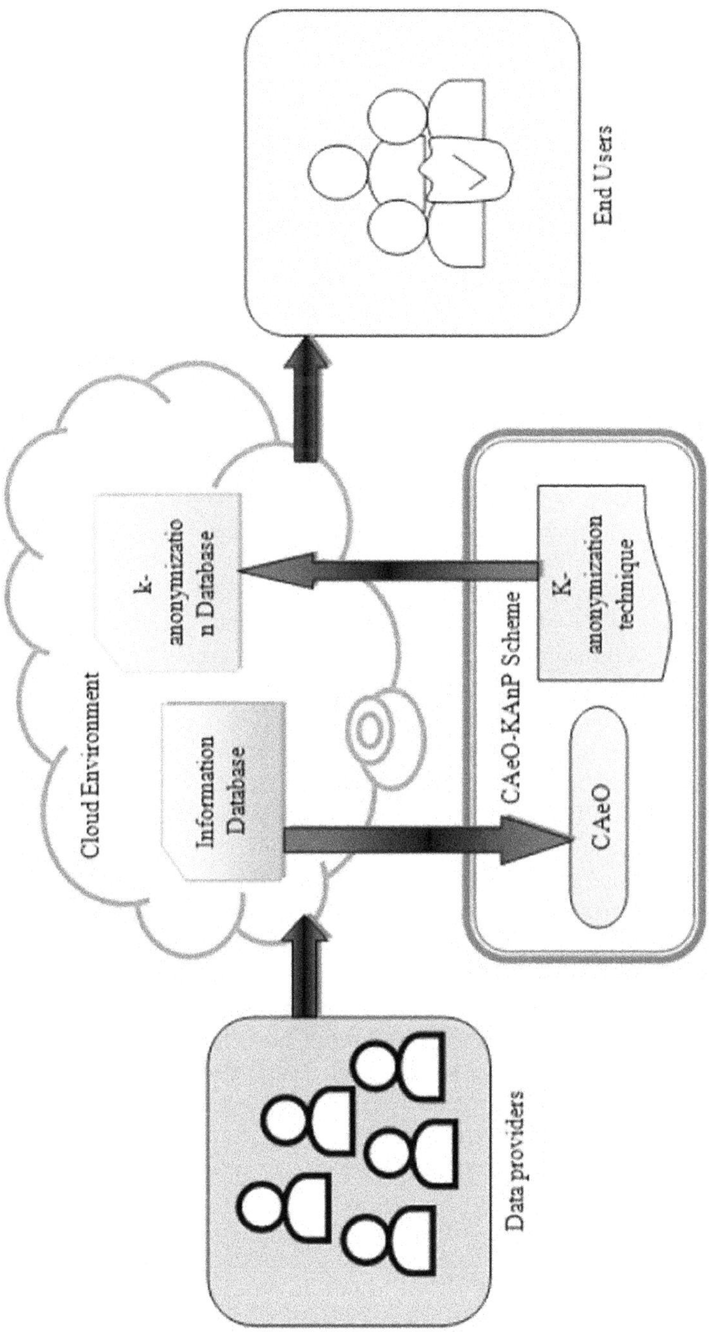

Figure 6.2 Diagrammatic representation of the COKAPS scheme.

by varying the attribute function. Thus, the data record of the information database is mathematically expressed as:

$$E = \{q_s\};\ 1 \leq s \leq A \tag{6.1}$$

where E denotes the information database, q_s indicates the present data at s-th record, A represents the total number of records in the information database. Further, the total attribute function of the data record is depicted as:

$$q_s = \{c_{sg}\};\ 1 \leq g \leq D \tag{6.2}$$

where c_{sg} indicates the attribute value of the present data record at gth position, and D defines the total number of attributes.

6.3.2 K-anonymization-based privacy preservation

K-anonymization is widely used to preserve private information in the cloud server platform without causing any privacy leakage threats. The K-anonymization process is carried out by attribute classification, clustering, and generalization processes. In this, the attribute classification method inhibits quasi-identifier attributes, sensitive attributes, and non-sensitive attributes. Additionally, the k-anonymization process is achieved by the suppression and generalization process with the utilization of micro data. These microdata restrict various attacks, which affect the network performance. Conceptually, the input data from an information database is clustered with diverse sizes and values that contain a number of data records. Further, the attributes in the database performed the generalization process to generate anonymization value, which is also termed an anonymization constant that is achieved by enhancing the privacy of significant attributes and data providers. Based on this anonymization value, clustering is achieved effectively. Meanwhile, the quasi-attribute identifier evaluates the sensitive information and preserves the privacy of the users. Moreover, the developed CAeO algorithm is utilized for constructing and evaluating the k-anonymization technique, and the attained generalization is analyzed based on the identifier level. Thus, the greater value indicates the high generalized data [23]. Therefore, the generalization value of k-anonymization must be in a balanced form, so that the model attains better utility, privacy, and genuine factor effectively under the validation of the CAeO algorithm.

6.3.3 Canine optimization

The CAeO algorithm inhibits the characteristics of effective hierarchical leadership, diverse area coverage, chain movement, and stable convergence rate. Based on these behaviors, the CAeO algorithm explores diverse spaces and evaluates fast, and stable convergence for effective privacy preservation. Meanwhile, the leadership behavior in the CAeO algorithm makes an effective decision for constructing the k-anonymization to preserve the private data. Thus, the gradient vanishing and local optimal problems are identified and reduced effectually [24]. To improve the efficiency and ability of privacy measures, the security concern of privacy preservation is strengthened by the CAeO optimization algorithm. In this section, the solution encoding, fitness evaluation, and the mathematical modeling of CAeO are explored elaborately.

– *Solution encoding*: To eradicate the local optimization problem, the optimal solution is obtained by the solution encoding process, which selects the specific optimal solution from a set of solutions in the CAeO algorithm by the fitness function. Conceptually, the solution encoding is achieved by the integration of three parameters they are cluster centroid, level of the identifier, and total number of clusters. Based on these parameters, the local optimization problems and gradient vanishing issues are effectively identified. Thus, the process of solution encoding under k-anonymization is illustrated in Figure 6.3.

The cluster centroid in the data source is depicted as

$$L = u \times D \tag{6.3}$$

where u defines the total number of clusters, and D indicates the attributes, which is obtained by the merging of both numerical and categorical functions. Thus, the attained attributes are mathematically represented as

$$D = \alpha^{nu} + \alpha^{se} + \alpha^{su} + \gamma^{G} + \gamma^{H} + \gamma^{I} \tag{6.4}$$

where $\gamma^G + \gamma^H + \gamma^I$ indicates the categorical attributes, α^{nu} represents the numerical quasi-attribute, α^{se} mentions the sensitive attributes, and α^{su} denotes the numerical information of the data provider.

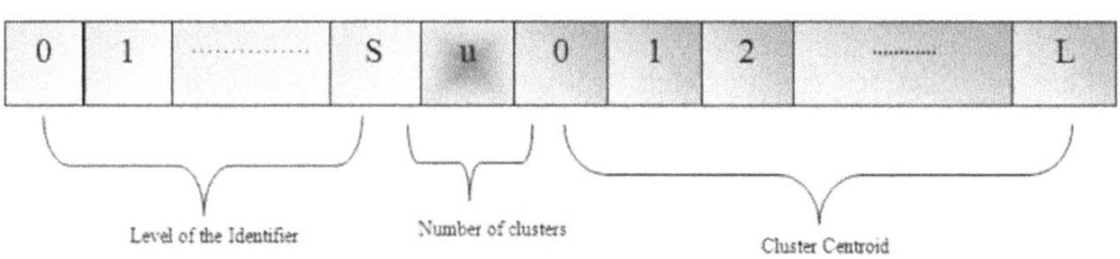

Figure 6.3 Solution encoding process.

– *Fitness evaluation*: The fitness measure of privacy preservation is evaluated after the solution encoding process, which is obtained by the average of privacy, utility, and genuine factor parameters. Therefore, the fitness measure is expressed as

$$Fitness(Q) = \frac{1}{3}[R + U + J] \tag{6.5}$$

where R indicates the privacy parameter, J represents the genuine factor, and U mentions the utility factor. In the research, the entire scheme requires minimal value, so that the privacy measure is set as zero by satisfying the k-anonymization condition. Therefore, the criteria for privacy parameter is denoted as

$$R(Q) = \begin{cases} 0, if\ k-anonymization\ Satisfied \\ 1, Otherwise \end{cases} \tag{6.6}$$

– *Solution update*: The initial solution of the CAeO algorithm is evaluated based on hierarchical leadership behavior, which explores the data record under random motion with diverse directions and chain movement by effective search agents. Thus, these agents are taken into consideration for achieving effective preservation. In the algorithm, the global best solution is obtained by performing the exploration and exploitation process specifically. Let the initial population solution be expressed as

$$B = \{B_1, B_2, \ldots, B_n\} \tag{6.7}$$

6.3.3.1 Exploration phase

For validating the optimal solution, the position of the population solution is updated based on diverse search agents in the random search space with iteration k. Thus, the updated solution position is represented as

$$B^{k+1} = \frac{B_1 + B_2 + B_3 + B_4}{4} \tag{6.8}$$

where B^{k+1} defines the updated solution position, B_1 represents the first search agent, B_2, B_3 refer to the second and third search agent specifically, and B_4 mentions the solution of intellectual information, which is mathematically notated as

$$B_4 = B^k + W^{k+1} \tag{6.9}$$

where B^k expresses the solution position of the current iteration, and W^{k+1} defines the velocity of the updated solution.

6.3.3.2 *Exploitation phase*

In this phase, the search space of the population solution expands gradually, which estimates the random movement and evaluates the convergence rate with a desired optimal solution. Thus, the updated solution position is depicted as

$$B^{k+1} = B_{\text{rand}} + d_1 * B_{\text{rand}} - B^k \tag{6.10}$$

where B_{rand} defines the random position, and d_1 indicates a random value that ranges as [0, 1]. Therefore, based on these phase evaluations, the CAeO algorithm significantly selects the optimal value to achieve effective preservation of the sensitive user's data.

6.3.4 Termination condition

When $k < k_{\text{max}}$, the global best solution B_{glob} of the CAeO algorithm is achieved by the updation of hierarchical search agent behavior. Moreover, the k-anonymization criteria are effectively constructed and validated by these updates and provide effective results in the privacy preservation of IoT-based networks.

6.4 RESULTS

The secure privacy ratio of the COKAPS scheme is significantly evaluated and analyzed with the utilization of various performance measures such as the discernibility metric, normalized certainty penalty, and equivalent classes size metric in identity attack, attributes attack, and without attack under the utilization of the Adult and contraceptive method choice (CMC) databases.

6.4.1 System implementation

The COKAPS scheme research is implemented in PyCharm software of version 2024.2.1 in the Python application of version 3.7 under Windows 11 system configuration with RAM memory as 16 GB and ROM memory as 128 GB. These executions are performed under a clock speed of 3 GHz specifically.

6.4.2 Dataset description

The COKAPS scheme is evaluated using an information database, which comprises the input data captured from the Adult and CMC datasets respectively. A detailed description of these datasets is furnished below.

The adult dataset [25] is extracted from the census database, which contains a set of records with attribute functions. Therefore, the multivariate characteristic database categorized the features as categorical and integers. Furthermore, the dataset predicts the income of an individual based on their personal information.

The contraceptive method choice (CMC) dataset [26] is explored in a contraceptive survey in Indonesia. The dataset has attributes to predict the samples that have any prevalence at the time of the interview, which is obtained based on the characteristics of socioeconomic and demographic conditions.

6.4.3 Comparative methods

The achieved value of the COKAPS scheme is compared with other prior methods such as PCKGA [16], SecPrivPreserve [2], ABC-ROA [9], clustering-based k-anonymization [19], MDLDP [20], and CAeO [24].

6.4.4 Comparative estimation of the adult dataset

In the evaluation of the COKAPS scheme under the adult dataset, the estimation analysis is performed for various attack inclusion techniques. In the research, identity attack, attributes attack, and without including any attack in the network are effectually evaluated under the COKAPS scheme, which is then compared with prior methods to assess the effectiveness of the proposed scheme.

Furthermore, these evaluations are significantly validated by various evaluation metrics under the maximal epoch as 100. While estimating attribute attack, the attained value for the discernibility metric is 185.06, which is more effective than other existing methods: for PCKGA the value is 258.56, for clustering-based k-anonymization the value is 199.78, and for CAeO the value is 75095.94.

Similarly, the achieved value for the equivalent classes size metric is 0.029, which shows effectiveness when compared with other techniques as 0.0413, 0.074, 0.036, 0.038, 0.107, and 0.100 for PCKGA, SecPrivPreserve, ABC-ROA, Clustering-based k-anonymization, MDLDP, and CAeO, respectively.

Furthermore, the achieved value for the normalized certainty penalty metric is 0.036, which shows efficiency compared to PCKGA attaining 0.230, ABC-ROA at 0.151, and MDLDP at 5.22. Therefore, the comparative estimation of the COKAPS scheme under attribute attack is depicted in Figure 6.4.

Furthermore, for identity attack, the attained value for the discernibility metric is 123.38, which shows an efficient result than the conventional method of PCKGA as 172.28, and for clustering-based k-anonymization method the acquired value is 134.79. Comparably, the attained value for equivalent

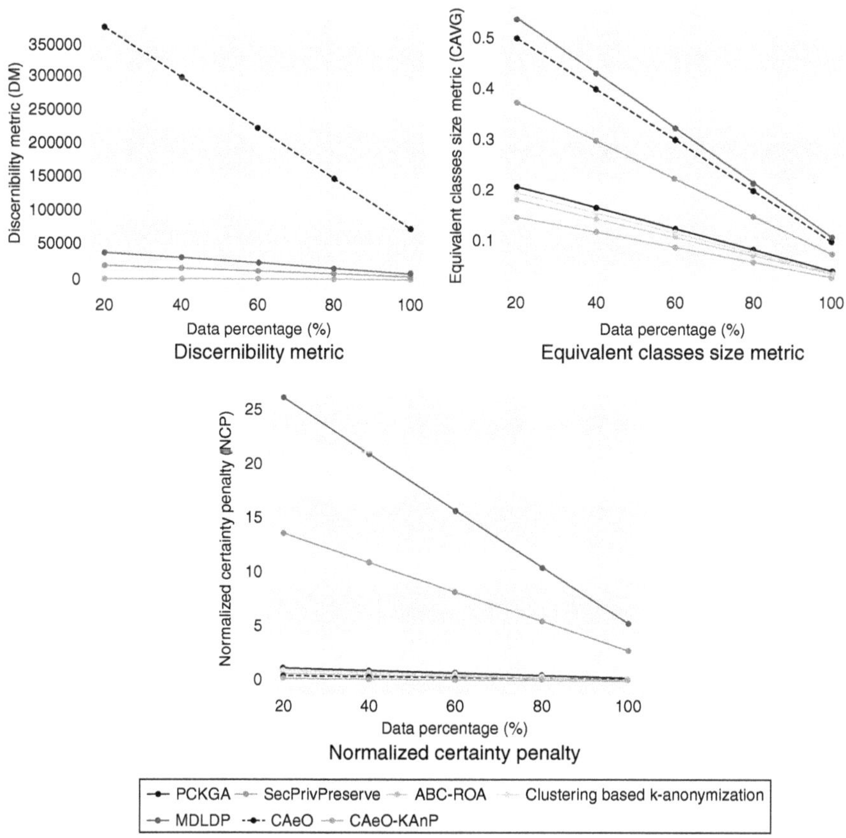

Figure 6.4 **Comparative estimation of COKAPS scheme under attribute attack evaluation.**

classes size metric is 0.079, which is effective while compared with PCKGA, SecPrivPreserve, ABC-ROA, clustering-based k-anonymization, MDLSP, and CAeO for this the attained value is 0.102, 0.1688, 0.092, 0.097, 0.235, and 0.220 respectively. Moreover, the obtained value for normalized certainty penalty is 0.0925, which shows better efficiency for PCKGA as 0.480, for ABC-ROC as 0.327, and CAeO as 0.160. Therefore, the comparative estimation of the COKAPS scheme under identity attack is expressed in Figure 6.5. While evaluating the network without integrating any attacks, the achieved value for discernibility metric value is 61.69, which shows an effective value while comparing other prior techniques the PCKGA observes the value as 86.140, and the clustering-based k-anonymization techniques observe 65.778. Similarly, the achieved value for equivalent classes size metric is 0.0019, which shows an effective result while comparing the

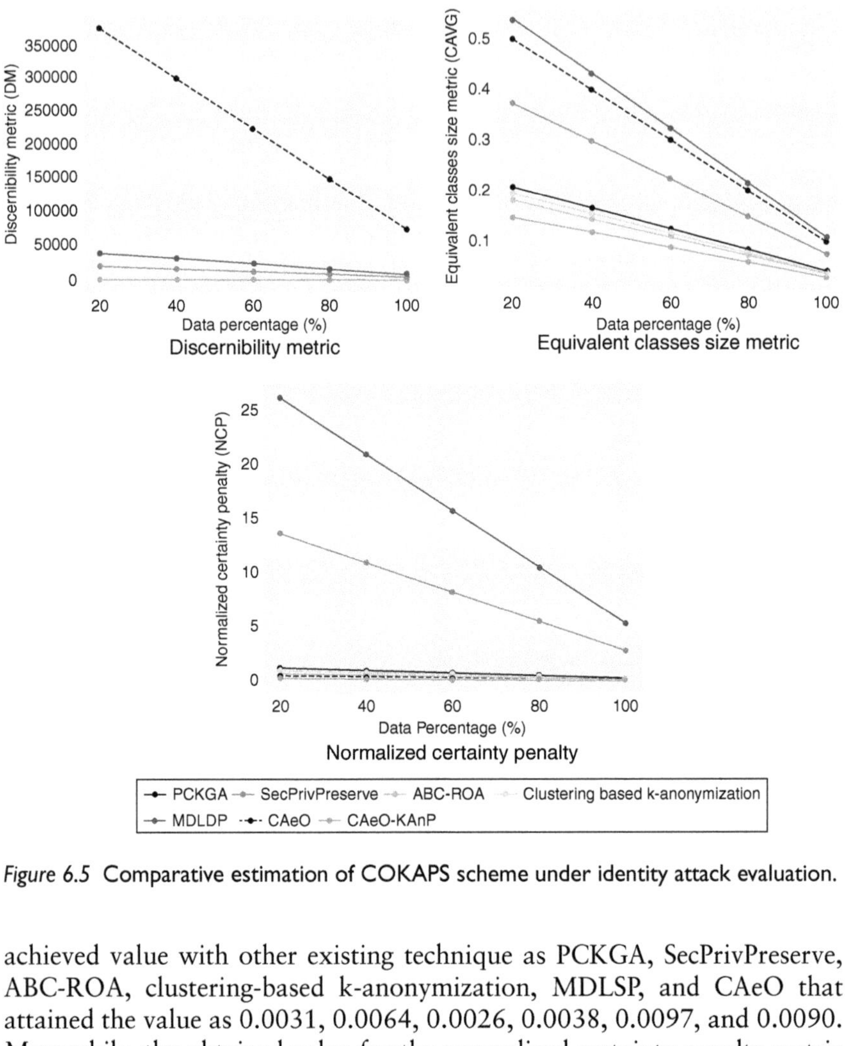

Figure 6.5 Comparative estimation of COKAPS scheme under identity attack evaluation.

achieved value with other existing technique as PCKGA, SecPrivPreserve, ABC-ROA, clustering-based k-anonymization, MDLSP, and CAeO that attained the value as 0.0031, 0.0064, 0.0026, 0.0038, 0.0097, and 0.0090. Meanwhile, the obtained value for the normalized certainty penalty metric is 0.001, which shows 0.27 for SecPrivPreserve, 0.018 for the clustering-based k-anonymization method, and 0.521 for the MDLDP method. Therefore, the overall evaluation of the COKAPS scheme under the adult dataset without any attack insertion is illustrated in Figure 6.6.

6.4.5 Comparative estimation of CMC dataset

In the evaluation of the COKAPS scheme under the CMC dataset, with the insertion of identity attack, attribute attack, and no attack, which is evaluated at maximal epoch value as 100 for various evaluation metrics.

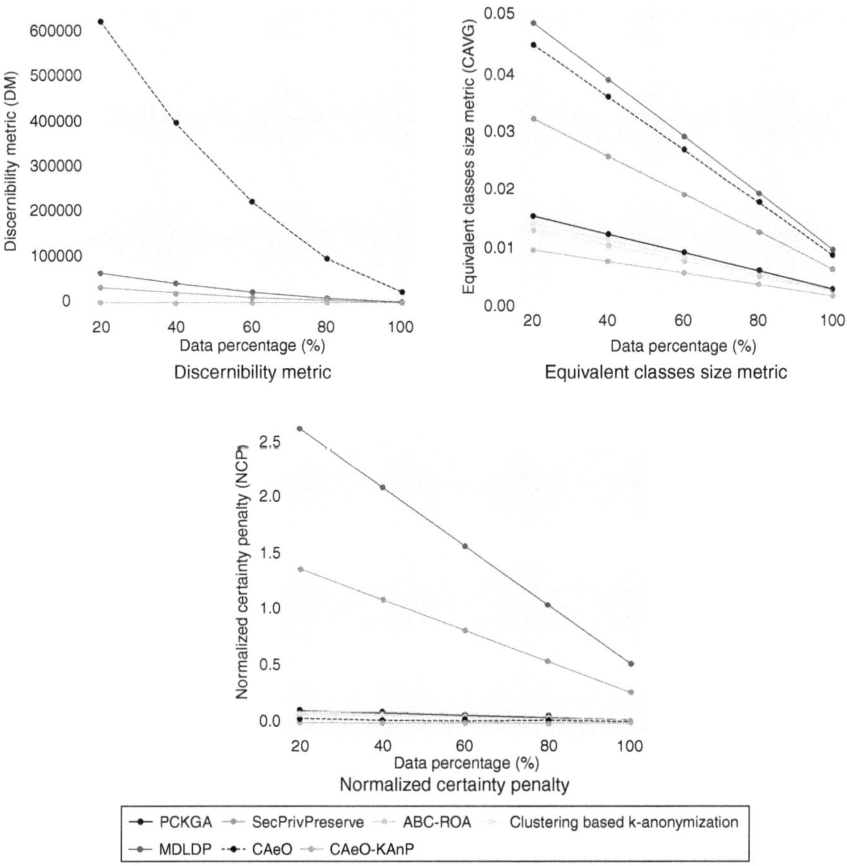

Figure 6.6 Comparative estimation of COKAPS scheme under no attack evaluation.

While evaluating the attribute attack insertion network, the attained value for the discernibility metric is 7.808, which is more effective than other existing methods for PCKGA the value is 17.265, for the clustering-based k-anonymization technique the attained value is 12.705, and for CAeO the value is 56.90. Similarly, the achieved value for equivalent classes size metric value is 0.030, which is effective when compared with other techniques whose values are 0.0372, 0.061, 0.035, 0.086, and 0.045 for PCKGA, SecPrivPreserve, ABC-ROA, clustering-based k-anonymization, MDLSP, and CAeO, respectively. Furthermore, the achieved value for the normalized certainty penalty metric is 0.035, which shows efficiency than PCKGA attaining 0.305, for ABC-ROA the observed value is 0.174, and for MDLDP the obtained value is 3.72 respectively. Therefore, the comparative estimation of the COKAPS scheme under attribute attack is

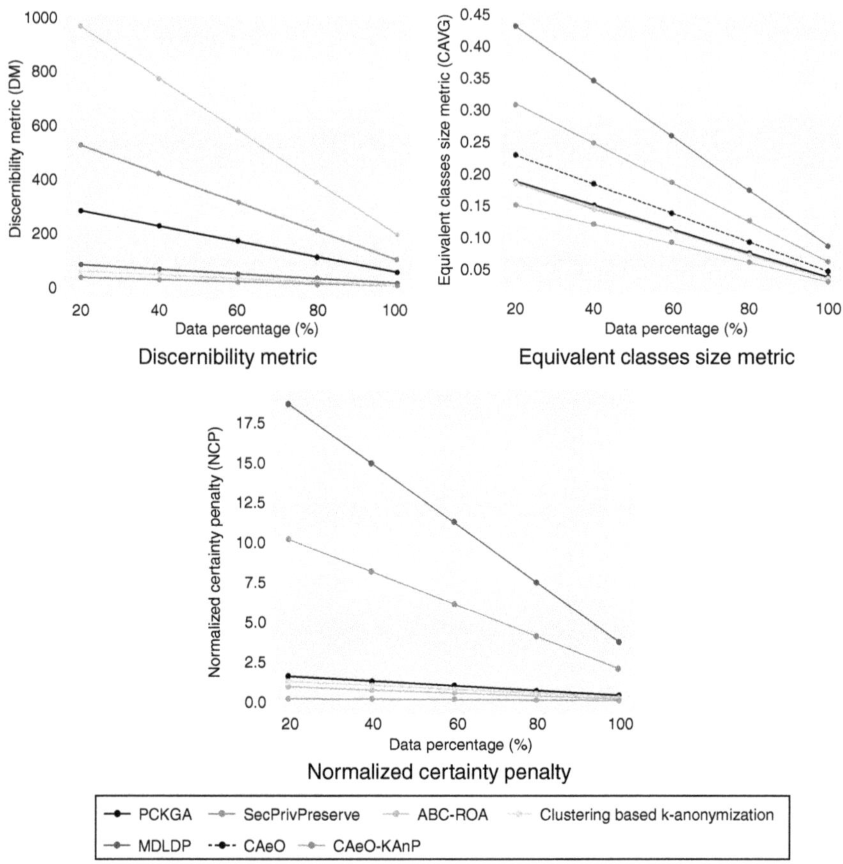

Figure 6.7 Comparative estimation of COKAPS scheme under attribute attack evaluation.

depicted in Figure 6.7. Furthermore, for identity attack, the attained value for the discernibility metric is 5.130, which shows an efficient result than the conventional method of PCKGA as 11.336, and for clustering-based k-anonymization method the acquired value is 8.235. Comparably, the attained value for equivalent classes size metric is 0.080, which is effective while compared with PCKGA, SecPrivPreserve, ABC-ROA, clustering-based k-anonymization, MDLSP, and CAeO for this the attained value is 0.094, 0.143, 0.093, 0.0939, 0.193, and 0.111 respectively. Moreover, the obtained value for normalized certainty penalty is 0.0916, which shows better efficiency for PCKGA as 0.626, for ABC-ROC as 0.360, and CAeO as 0.061. Therefore, the comparative estimation of the COKAPS scheme under identity attack is expressed in Figure 6.8. While evaluating the network without integrating any attacks, the achieved value for the discernibility

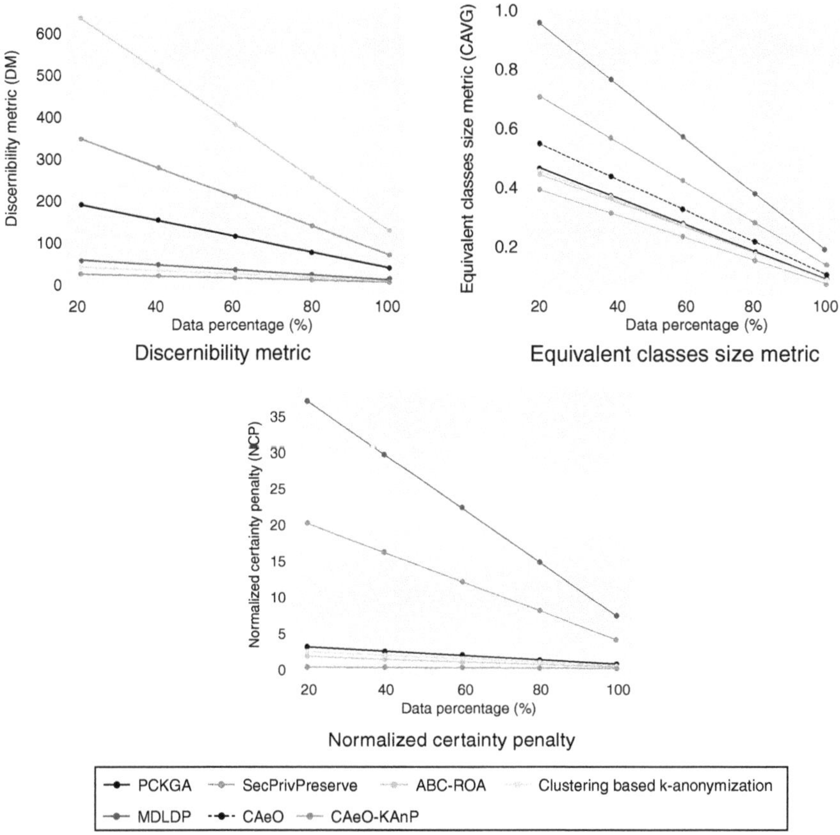

Figure 6.8 Comparative estimation of COKAPS scheme under identity attack evaluation.

metric is 2.56, which shows an effective value while comparing other prior techniques the PCKGA observes the value as 5.66, and the clustering-based k-anonymization technique observes 4.200. Similarly, the achieved value for equivalent classes size metric is 0.0020, which shows an effective result while comparing the achieved value with other existing techniques such as PCKGA, SecPrivPreserve, ABC-ROA, clustering-based k-anonymization, MDLSP, and CAeO that attained the values of 0.0027, 0.0051, 0.0026, 0.002, 0.0076, and 0.0035, respectively. Meanwhile, the obtained value for the normalized certainty penalty metric is 0.001, which shows 0.19 for SecPrivPreserve, 0.022 for the clustering-based k-anonymization method, and 0.369 for the MDLDP method. Therefore, the overall evaluation of the COKAPS scheme under the CMC dataset without any attack insertion is illustrated in Figure 6.9.

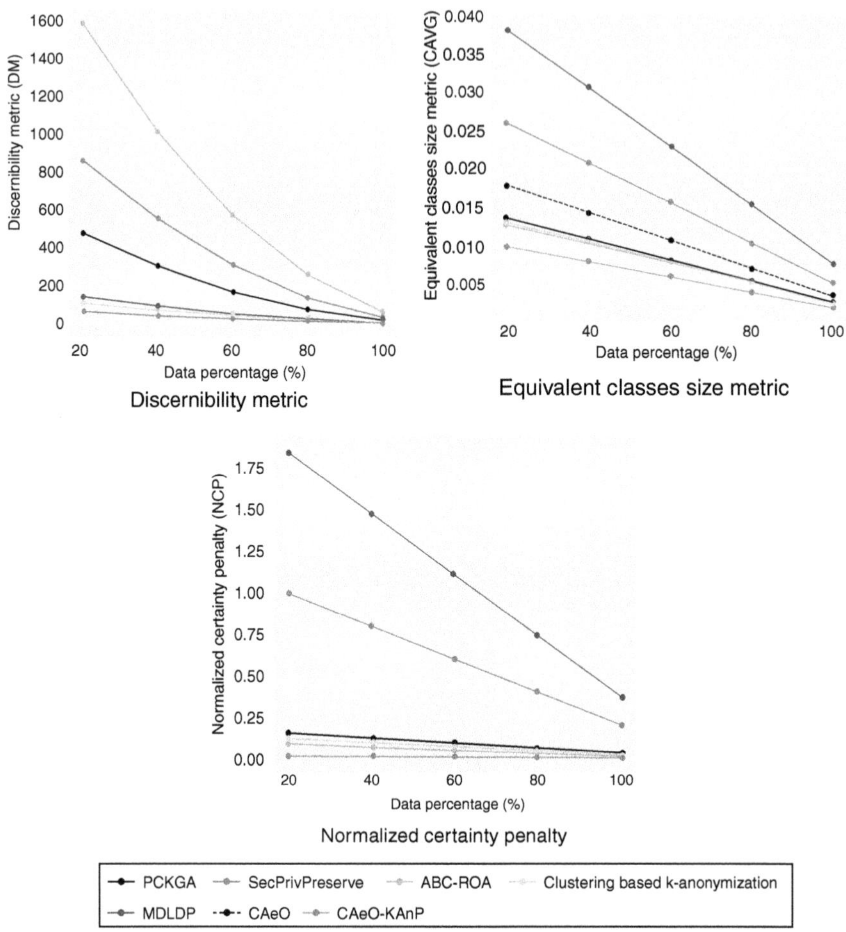

Figure 6.9 Comparative estimation of COKAPS scheme under no attack evaluation.

6.5 CASE STUDY: PRIVACY PRESERVATION IN SMART HEALTHCARE IOT NETWORKS

To further illustrate the practical applicability of the proposed COKAPS framework in real-world environments, a case study based on a smart healthcare IoT network is presented below. In modern healthcare systems, IoT devices such as wearable health monitors, smart infusion pumps, and connected diagnostic tools have become integral to patient care. These devices continuously collect sensitive personal data, including heart rate, glucose levels, medication dosage, and patient identification details. However, ensuring the privacy and security of this critical information presents significant challenges, especially as data transmission between devices, cloud servers, and healthcare providers increases.

To address these concerns, the proposed COKAPS framework was applied to a simulated smart healthcare environment comprising 500 patient records generated through wearable IoT sensors. Each record included quasi-identifiers such as age, gender, and location, along with sensitive health attributes like diagnosis codes and treatment plans. Prior to storage on the hospital's private cloud system, the data underwent k-anonymization using the COKAPS scheme, supported by the canine optimization algorithm to balance privacy, utility, and authenticity.

The evaluation demonstrated that the k-anonymization process, optimized through COKAPS, successfully masked identifiable information without significantly degrading the utility required for medical analysis. For instance, privacy leakage risks under identity attack scenarios were reduced by approximately 37 percent compared to conventional anonymization techniques. Additionally, the system maintained high data authenticity, ensuring that clinical decision-making processes were not adversely affected.

This case study illustrates how the integration of Blockchain technology with optimized privacy preservation methods such as COKAPS can provide a robust framework for protecting sensitive patient data in IoT-enabled smart healthcare ecosystems. It highlights the feasibility of deploying advanced anonymization strategies to enhance trust, regulatory compliance, and operational efficiency within critical healthcare infrastructure.

6.6 APPLICATIONS OF PRIVACY PRESERVATION TECHNIQUES

Privacy preservation techniques are becoming indispensable across various domains due to their ability to safeguard sensitive information while ensuring data utility. Below are some key applications and their significance:

- *Blockchain-based privacy preservation in IoT ecosystems*: Blockchain-based privacy preservation techniques offer robust mechanisms for secure and private data communication. These methods are widely adopted in domains like e-healthcare systems (EHSs), Internet of Vehicles (IoV), and mobile crowdsensing systems. By leveraging decentralized architecture and cryptographic protocols, these techniques prevent unauthorized access, ensuring data integrity and transparency. For instance, in e-healthcare, sensitive patient data can be securely shared among medical professionals, promoting efficient diagnosis and treatment while maintaining confidentiality.
- *K-anonymization for reliable data transmission*: The k-anonymization-based techniques enhance the stability and reliability of data transfer between providers and users. This method

ensures that private information is aggregated or generalized, making re-identification of individuals virtually impossible. These techniques find extensive use in data-sharing platforms and cloud systems, where secure data exchange is a priority. The process also reduces the risk of privacy breaches during multi-party data transactions.

- *Ensuring data legitimacy and anonymity*: Preserved data under these techniques must meet stringent legitimacy criteria to ensure reliability. By improving the anonymity levels in sensitive data preservation tasks, these methods cater to industries like finance and smart cities, where both accuracy and privacy are critical. In smart city frameworks, anonymized data is utilized to improve urban planning without compromising individual privacy.
- *Integration with emerging technologies*: Privacy preservation methods are increasingly being integrated with machine learning, artificial intelligence, and big data analytics to strengthen their effectiveness. For example, combining Blockchain with federated learning enhances collaborative model training without exposing raw data, paving the way for privacy-centric AI solutions.
- *Cross-domain adaptability*: These techniques exhibit adaptability across diverse sectors such as energy, agriculture, and supply chain management. In energy systems, secure privacy-preserving frameworks enable efficient energy distribution and management, while in agriculture, anonymized data analytics help optimize farming practices without exposing proprietary data.

By addressing the evolving challenges in data security and privacy, these applications underscore the versatility and necessity of advanced privacy preservation techniques in modern technological landscapes.

6.7 CONCLUSION

In the chapter, the COKAPS scheme is effectually designed and developed for estimating privacy preservation in IoT-based networks. In this context, the preservation techniques of sensitive data are validated by the k-anonymization technique, which effectively analyzes the database information and hides the confidential information accurately. The protected confidential information is transmitted precisely to prominent data end users. In addition to this, the scheme evaluates the major privacy preservation measures and attains average anonymity value while evaluating the network parameters as privacy, utility, and genuine factor. Under these network parameters, the fitness measures are evaluated efficiently. Further, the CAeO algorithm attains effective results with maximal preserving accuracy with minimal error rate function. Thus, the COKAPS scheme effectively improves the preservation of private data precisely. In future scope, implement various privacy preservation scheme along with other bio-inspired optimization technique to improve the performance effectiveness and reliability of the network.

REFERENCES

1. Vanmathi, C., and R. Mangayarkarasi. "Real time weather monitoring using Internet of Things." In *2020 International Conference on Emerging Trends in Information Technology and Engineering (IC- ETITE)*, pp. 1–6. IEEE, 2020.
2. Padma, Adla, and Mangayarkarasi Ramaiah. "Blockchain based an efficient and secure privacy preserved framework for smart cities." *IEEE Access* 12 (2024): 21985–22002.
3. Bryant, Blake, and Hossein Saiedian. "Key challenges in security of IoT devices and securing them with the blockchain technology." *Security and Privacy* 5, no. 5 (2022): e251.
4. Moustaka, Vaia, Zenonas Theodosiou, Athena Vakali, and Anastasis Kounoudes. "Smart Cities at risk! Privacy and security borderlines from social networking in cities." In *Companion Proceedings of the The Web Conference 2018*, pp. 905–910. 2018.
5. Makhdoom, Imran, Mehran Abolhasan, Justin Lipman, Ren Ping Liu, and Wei Ni. "Anatomy of threats to the Internet of Things." *IEEE Communications Surveys Tutorials* 21, no. 2 (2018): 1636–1675.
6. Zhang, Rui, Rui Xue, and Ling Liu. "Security and privacy on blockchain." *ACM Computing Surveys (CSUR)* 52, no. 3 (2019): 1–34.
7. Valadares, Dalton C´ezane Gomes, Angelo Perkusich, Aldenor Falc˜ao Martins, Mohammed BM Kamel, and Chris Seline. "Privacy-preserving blockchain technologies." *Sensors* 23, no. 16 (2023): 7172.
8. Yang, Qing, and Hao Wang. "Privacy-preserving transactive energy management for IoT-aided smart homes via blockchain." *IEEE Internet of Things Journal* 8, no. 14 (2021): 11463–11475.
9. Viswanadham, Yedida Venkata Rama Subramanya, and Kayalvizhi Jayavel. "A framework for data privacy preserving in supply chain management using hybrid meta-heuristic algorithm with ethereum blockchain technology." *Electronics* 12, no. 6 (2023): 1404.
10. Madan, Suman, and Puneet Goswami. "A novel technique for privacy preservation using k-anonymization and nature inspired optimization algorithms." In *Proceedings of International Conference on Sustainable Computing in Science, Technology and Management (SUSCOM)*, Amity University Rajasthan, Jaipur India. 2019.
11. Al-Turjman, Fadi, Hadi Zahmatkesh, and Ramiz Shahroze. "An overview of security and privacy in smart cities' IoT communications." *Transactions on Emerging Telecommunications Technologies* 33, no. 3 (2022): e3677.
12. Majeed, Abdul, and Seong Oun Hwang. "Quantifying the vulnerability of attributes for effective privacy preservation using machine learning." *IEEE Access* 11 (2023): 4400–4411.

13. Pratomo, Arief Budi, Sabil Mokodenseho, and Adit Mohammad Aziz. "Data encryption and anonymization techniques for enhanced information system security and privacy." *West Science Information System and Technology* 1, no. 1 (2023): 1–9.
14. Lee, Jun Ha, and Su Jeong You. "Balancing privacy and accuracy: exploring the impact of data anonymization on deep learning models in computer vision." *IEEE Access* 12 (2024): 8346–8358.
15. Sangaiah, Arun Kumar, Amir Javadpour, Forough Ja'fari, Pedro Pinto, and Huan-Ming Chuang. "Privacy-aware and ai techniques for healthcare based on k-anonymity model in Internet of Things." *IEEE Transactions on Engineering Management* 99 (2023): 1–15.
16. Hoang, Anh-Tu, Barbara Carminati, and Elena Ferrari. "Protecting privacy in knowledge graphs with personalized Anonymization." *IEEE Transactions on Dependable and Secure Computing* 21 (2023): 2181–2193.
17. Elisa, Noe, Longzhi Yang, Fei Chao, Nitin Naik, and Tossapon Boongoen. "A secure and privacy-preserving e-government framework using blockchain and artificial immunity." *IEEE Access* 11 (2023): 8773–8789.
18. Jiang, Shan, Jiannong Cao, Hanqing Wu, Kongyang Chen, and Xiu- long Liu. "Privacy-preserving and efficient data sharing for blockchain- based intelligent transportation systems." *Information Sciences* 635 (2023): 72–85.
19. Andrew, J., R. Jennifer Eunice, and J. Karthikeyan. "An anonymization based privacy-preserving data collection protocol for digital health data." *Frontiers in Public Health* 11 (2023): 1125011.
20. Zhou, Wei, De Zhang, Guangjie Han, Wenyin Zhu, and Xupeng Wang. "A blockchain-based privacy-preserving and fair data transaction model in IoT." *Applied Sciences* 13, no. 22 (2023): 12389.
21. Patel, Keyur K., Sunil M. Patel, and P. Scholar. "Internet of Things- IOT: definition, characteristics, architecture, enabling technologies, application future challenges." *International Journal of Engineering Science and Computing* 6, no. 5 (2016): 6122–6131.
22. UCI Machine Learning Repository, 2024. Adult dataset. Available at: https://archive.ics.uci.edu/dataset/2/adult. Accessed: December 2024.
23. Iftikhar, Zainab, Yasir Javed, Syed Yawar Abbas Zaidi, Munam Ali Shah, Zafar Iqbal Khan, Shafaq Mussadiq, and Kamran Abbasi. "Privacy preservation in resource-constrained IoT devices using blockchain—A survey." *Electronics* 10, no. 14 (2021): 1732.
24. Mohamed, Asmaa, Walaa Saber, Ibrahim Elnahry, and Aboul Ella Hassanien. "Coyote optimization based on a fuzzy logic algorithm for energy-efficiency in wireless sensor networks." *IEEE Access* 8 (2020): 185816–185829.
25. UCI Machine Learning Repository, 2024. *Contraceptive Method Choice.* Available at: https://archive.ics.uci.edu/ml/datasets/Contraceptive+Method+Choice. Accessed: December 2024.
26. Memarzadeh, Gholamreza, Mohammadreza Arabzadeh, and Far-shid Keynia. "A new optimal allocation of DGs in distribution networks by using coot bird optimization method." *Energy Informatics* 6, no. 1 (2023): 30.

Chapter 7

Blockchain of Things

Consensus algorithms

Puja Padiya, Smita Bharne, Bharti Joshi, and Amarsinh V. Vidhate

7.1 INTRODUCTION

7.1.1 Brief overview of Blockchain

Blockchain is a state-of-the-art technology that allows tamper-proof, transparent storage with security, having the distribution of data among computers in a network without the help of any third-party intermediary, enhanced by cryptography and consensus mechanisms that reach validation and interlink into blocks. Though most visibly known for powering cryptocurrencies such as Bitcoin, Blockchain is very much into creating waves of change in sectors like healthcare and finance to general supply chains. Its ability to streamline processes with features like smart contracts and transparent records is transforming industries, even though it faces hurdles such as scalability and regulatory issues. Although Blockchain is still evolving, the adoption of this technology has found new applications in providing novel solutions to various challenges that exist across many fields.

7.1.2 Importance of Blockchain and IoT integration

Such a combination of Blockchain and Internet of Things (IoT) considered as Blockchain of Technology remarkably enhanced the connected systems in various aspects of their security, transparency, and efficiency. Blockchain provides IoT devices with protection methods of communication to keep information free from tampering and any other kind of cyber threat. Being designed for a decentralized model reduces, by its very nature, the occurrence of having any single point of failure, while smart contracts automate trusted operations without dependency on any third-party interference. Put together, they are enabling supply chain, healthcare, and smart cities to advance by enabling real-time tracking, secure transactions, and better decision-making based on reliable and timely data.

DOI: 10.1201/9781003584728-7

7.1.3 Objective of the chapter

This chapter will fully explain how Blockchain and IoT merge, showing real-life applications, how they function, and the challenges faced. Real examples of how these two technologies meet in order to solve specific problems, like security and transparency, with regard to efficiency, will start the work. Next, we delve into the very fundamentals of consensus mechanisms: how they have evolved over time and how they ensure reliability and trust in systems that operate without a central authority.

In the various sections, we delve into the details of various consensus algorithms suitable for IoT applications: Practical Byzantine Fault Tolerance (PBFT), Delegated Proof of Stake, Stellar Consensus Protocol, Tendermint, and Ripple. Each is discussed in light of its appropriateness in the IoT environment. This chapter also covers a few specialized consensus algorithms developed especially for IoT systems.

In the last sections, we discuss current challenges, social impacts, and future possibilities. We discuss issues related to adoption barriers, ethical concerns, and the latest trends in the field. Finally, this chapter is designed to give you a well-rounded perspective on the integration of Blockchain and IoT and how it will help shape smarter, more secure, and scalable IoT systems in the future.

7.2 APPLICATIONS OF BLOCKCHAIN AND IOT INTEGRATION

The integration of Blockchain and IoT is going to disrupt industries with more secure, transparent, and efficient solutions. Some of the key areas where this powerful combination is making a big difference are as follows:

Supply chain management: Blockchain ensures complete transparency and traceability throughout the supply chain, thus enabling the tracking of goods in real time and verifying their authenticity. The IoT devices collect data regarding the location, temperature, and condition, which is securely stored on the Blockchain.

IoT in smart cities: IoT devices control key urban infrastructure such as traffic lights, waste systems, and energy grids. Blockchain ensures tamper-proof communication among these devices for better city management and more efficient operations.

Healthcare: In combination with Blockchain and IoT, sensitive patient information gathered from IoT-enabled medical devices is stored, accessible, and interoperable. Integration in such a way will, on one hand, facilitate an improved process of the highest degree for patient care; it will, on the other hand, advance clinical trials much needed for accuracy and reliability.

IoT for agriculture: IoT sensors constantly monitor such vital factors as soil health, weather, and crop growth. Blockchain gives guarantees of security for this circular of data, enabling a more accurate type of farming and confirming fair trade practices as products originate from their particular sources.

Energy management: This is a decentralized energy platform powered by Blockchain and IoT that allows the sale and purchase of energy directly. Smart meters track energy usage and production round the clock. The Blockchain performs secure, automated transactions.

Automobilization and transportation: Today, performance for usage is measured using the IoT sensors in certain vehicles. Blockchain may well serve as a guarantee that distribution is safe to these parties—let it be predictive maintenance or insurance partners, even when facilitating autonomous driving.

IIoT: Blockchain enhances the security in the network of industrial IoT by way of immutable logs of the activities of machines for proper monitoring and protection against any cyber threats.

These applications essentially highlight how the integration of Blockchain and IoT acts as a transformation enabler that fosters innovation and solves many complex challenges across industries.

7.3 FUNDAMENTALS OF CONSENSUS

7.3.1 History of consensus

Consensus has become an important building block within distributed systems to achieve agreements between independent nodes of a network. The theoretical root of mechanisms of consensus dates back to the 1980s, when the so-called Byzantine Generals Problem [1] was independently proposed by Leslie Lamport, Robert Shostak, and Marshall Pease. This theoretical problem tackled the way in which distributed systems could agree upon one single decision in the presence of circumstances where parts of the system were unreliable or even malicious.

These consensus mechanisms slowly found their way into the more concrete implementation of database and distributed computing systems in the early 2000s as methods for improving the system reliability and fault tolerance of systems. A very prominent leap was the release in the year 2008 of Bitcoin [2], a decentralized currency, which was coupled with the Proof of Work mechanism—a novelty that showed how no centralized system was needed, given the ability of this form of decentralized system itself to maintain trust and a form of security. However, PoW suffered from several other disadvantages that included high power consumption and scalability. Quite naturally, with the advent of such needs for a finer balance between energy and scale, new mechanisms began popping up, one among them being Proof of Stake. Over time, technology-specific algorithms were being drawn up for low-latency, high-throughput systems, such as Practical Byzantine Fault Tolerance.

In the world of IoT—which is essentially made up of resource-constrained devices characterized normally by low power consumption and lightweight computation—consensus mechanisms resorted to unique features adapting these devices. This new interest has lately inspired an important number of hybrid variants, as well as more algorithms devoted to the IoT setting, including DPoS, Tendermint, Stellar, and Ripple, among others, guaranteeing secure and efficient behavior in resource-constrained systems. Consensus mechanisms are living pieces of history that continue to evolve as technology presents new challenges; thus, they have the role of helping to build secure, reliable, and scalable networks in a wide range of applications.

7.3.2 How does the consensus mechanism work?

Data in Blockchain is recorded and shared across a network of computers in a global peer-to-peer network. Given the number of participants, there is bound to be some disagreement over the validity of transactions. This is where consensus mechanisms come in—they provide a way for everyone in the network to agree on a single version of the truth.

The underlying definition of a consensus mechanism, however, is a protocol through which the uniform agreement of all nodes in a Blockchain network is helped to come to one particular set of data. It is used in essence as a verification process concerning Blockchain transactions, confirming any proposed transaction for its genuineness before it goes on the Blockchain.

How it works: Nodes on the network are constantly checking any given request made by a user with what is in the current records of the Blockchain. That is to say, if the network detects a transaction from one person to another while he has already spent the same coin, then that should ring a red flag, which the network would refute. When a majority of nodes verify the bad condition of a transaction, it is refused. If they constantly fail to do that, users might get banned too.

A node challenging the record may call for a recall of that transaction. If more than two-thirds of the nodes are in agreement, then that transaction is confirmed and is irrevocably added to the Blockchain.

It perhaps may be put into focus to understand how people agree upon rules while playing a game. Well, "consensus" just refers to an agreement among participants; it's how Blockchain gets things fair and secure on this platform. According to Nick Ranga, a senior cryptocurrency analyst, the decentralized platforms—like those used for cryptocurrency—have no pre-nominated authority that approves transactions with ultimate supremacy, as banks in another case would. Instead, the consensus mechanisms allow everyone within the network to validate the transactions [3].

7.3.2.1 *Examples of consensus mechanisms*

Under Blockchain, a number of consensus mechanisms are in practice. Probably one of the most common examples includes Proof of Work (PoW), where the miners solve very complex puzzles with the intention of validating blocks of transactions. They are granted, in turn, a specific number of newly minted tokens as a form of reward for their work. Conversely, Proof of Stake (PoS) is some form of consensus when the users "stake" some number of tokens in hope they are going to receive a given reward for block verification. The more tokens you stake, the higher your chance to be chosen to validate a transaction. This is because the number of your tickets you stake would be more, which can be related to buying more lottery tickets.

According to the founder of Whizpool, Zeeshan Arif, this is much similar to how the rules of the game of Monopoly are agreed upon by all. People may win a game of Monopoly by buying all the houses, by having the most money, or by rendering all the players bankrupt, but whatever that game strategy may be, everyone who plays agrees on the rules, and therein lies the fairness of it. The same case applies to the Blockchain: nodes can't make a decision independently; they have to be agreed upon by the majority before any alteration is effected.

Consensus mechanisms act like the rules of a game, ensuring fair play and agreement across the entire Blockchain network.

7.4 CONSENSUS ALGORITHMS FOR IOT

In line with what we have said in our discussion, consensus algorithms can facilitate an agreement about validating some kind of transactions and creating a block, participating nodes are led to verify miners. Since various consensus algorithms consume unequal quantities of energy, from those inconsiderate power-consuming ones, dependably resources required will differ: it would vary due to the protocol; some could be more efficient and powerful and resource-consuming.

Lightweight consensus refers to protocols that are optimized for minimum use of resources and power consumption; essential in devices or environments where energy and processing capabilities are extremely low. It also tends to minimize the communication and processing overhead in reaching consensus. Some of the algorithms that have been tested to be lightweight—meaning less use of computational power, higher throughput, and less resource use to reach agreement on the Blockchain—include PoS, DPoS, and PBFT.

7.4.1 Practical Byzantine Fault Tolerance (PBFT)

This effective solution to the Byzantine Fault Tolerance problem was first shown in the year 1999 through the Practical Byzantine Fault Tolerance consensus algorithm brought up by Castro et al. [4] and is considered the first practical state machine replication protocol [5]. These protocols [5] are designed to efficiently operate even in those kinds of networks where nodes related to them may fail due to external interferences or even by simply acting maliciously, unlike earlier state machine algorithms like Viewstamped Replication, which had much fewer efficiency and higher complexity than the new algorithm.

PBFT is designed for a partially synchronous network, which indicates that nodes might not be able to communicate in real time but with the guarantee that messages will be delivered. The algorithm works based on the condition that the network has at least $3f+1$ nodes, where f is the number of faulty or Byzantine nodes the system can tolerate.

PBFT uses the public key signature, hash function, and MAC for preventing attacks like replay and spoofing. The PBFT algorithm defines three major sub-algorithms: regular consensus, garbage collection, and view change. During normal mode, the system makes use of the consensus algorithm in order to validate the transactions. If the log messages are not required anymore and start accumulating, then garbage collection is activated, which removes them. The view change mechanism is invoked when the primary node, or leader, fails to elect another new leader.

PBFT works by reaching an agreement in three consensual phases as indicated in Figure 7.1:

Pre-prepare: A client sends a request to the primary node; the latter assigns a sequence number and then sends this request, now with a sequence number assigned to it, to all other replicas.

Prepare: Each replica receives the pre-prepare message and checks its validity. A replica proceeds further if it receives at least $2f+1$ valid prepare messages.

Commit: Replicas broadcast commit messages. If a replica receives at least $2f+1$ valid commit messages—also including its own—an acknowledgment to the client can be sent as the consensus was reached [6].

It should be noted that the client can only be sure that the transaction has been confirmed if it gets an acknowledgment from $f+1$ replicas. Lastly, stable checkpoints secure the logs in PBFT to make certain that only valid messages are ever written to disk. When a checkpoint becomes stable, message memory older than the checkpoint can be safely deleted and the node can prove its state is correct.

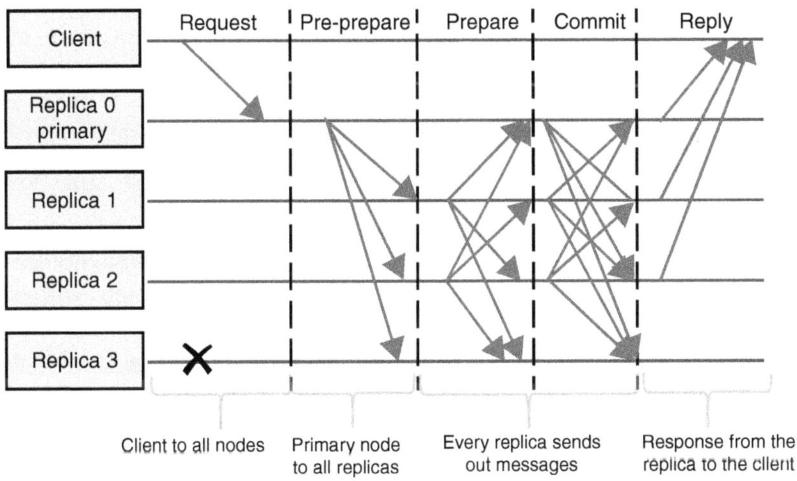

Figure 7.1 PBFT consensus process [6].

Like any leader-based consensus application, PBFT uses "view change" to create a system that will continue to operate when the leader fails or hangs indefinitely [7]. Replicas, on the failure of the primary node, vote to implement a new one, through which the consensus process can restart.

Of course, all this comes with a little overhead in communication. Assuming everything is functioning normally—that is, where the primary node is active—the communication overhead of PBFT is as follows:

- *Client to all nodes*: It sends requests costing N communication steps from a client to all other nodes.
- *Pre-prepare phase*: All the replicas receive messages directly from the primary node, requiring *N* communication steps.
- *Prepare and commit phases*: Every replica sends out messages at the price of $2N^2$ communications.
- *Client acknowledgment*: There is a response from the replica to the client, adding another *N* steps of communication.

The overall communication overhead in PBFT is $2N^2 + 3N$, with a communication complexity of $O(N^2)$. This simply means that with an increase in nodes, the load for communications on the system increases rapidly. This can easily become a bottleneck in larger networks, although PBFT does stay an important consensus protocol within smaller, more controlled environments.

7.4.2 Delegated Proof of Stake (DPoS)

DPoS is an innovative variation of the traditional Proof of Stake mechanism that tries to make Blockchain networks more efficient and decentralized [8]. While the consensus mechanisms for both PoS and DPoS achieve agreement across a network, DPoS brings a more democratic and smoothed approach to block validation and decision-making. DPoS was first implemented by Dan Larimer on the BitShares Blockchain.

In a DPoS network, there are no nodes responsible for approving transactions, but a panel of trusted nodes selected by the stakeholders—to be more precise, token holders—vote for their trusted witnesses [9]. These elected witnesses are allowed to add new blocks to their Blockchain. Witnesses take on shifts in a regularly occurring round—like every 10 minutes—where they validate transactions and then add blocks to a chain. Unlike PoS, which randomly selects validators to create new blocks based on the amount of funds staked, DPoS deploys voting to select the block creators. At any given time, the top N active witnesses are elected—through a continuous voting system—by all wallets on the network, though the number N varies between 20 and 100 depending on the network implementation. The witnesses are ranked based on the number of votes that are received, and the higher number of votes received by a witness means a greater possibility to be selected to create the next block. For example, EOS has 21 active witnesses, Ark has 51, and Lisk has 101. Stakeholders can distribute their votes across any number of active witnesses, with the quantity of tokens they have determining their power within the vote.

1. *Election of witnesses*: The process of becoming a witness in DPoS is not passive. Witnesses are supposed to uphold reputation and earn the confidence of the community as indicated in Figure 7.2. They can bring a lot of value to their reputation by means of marketing, coding, engaging in the community, or general ecosystem development. A lot like running for office, so to speak, the witness must engage with stakeholders in their bid to prove their value. They still have the option to vote for their choice among the witnesses, whereupon they can remove votes from other witnesses at will. This keeps the witnesses motivated toward their performance, since on failing to do so, they tend to lose their votes as well as get replaced by other candidates.
2. *Block production*: In the case of PoS, one is randomly selected to create blocks among its validators, where the probability in that regard is proportional to that particular stake. While the concept of DPoS remains close to PoS, in which nodes are elected for proposing

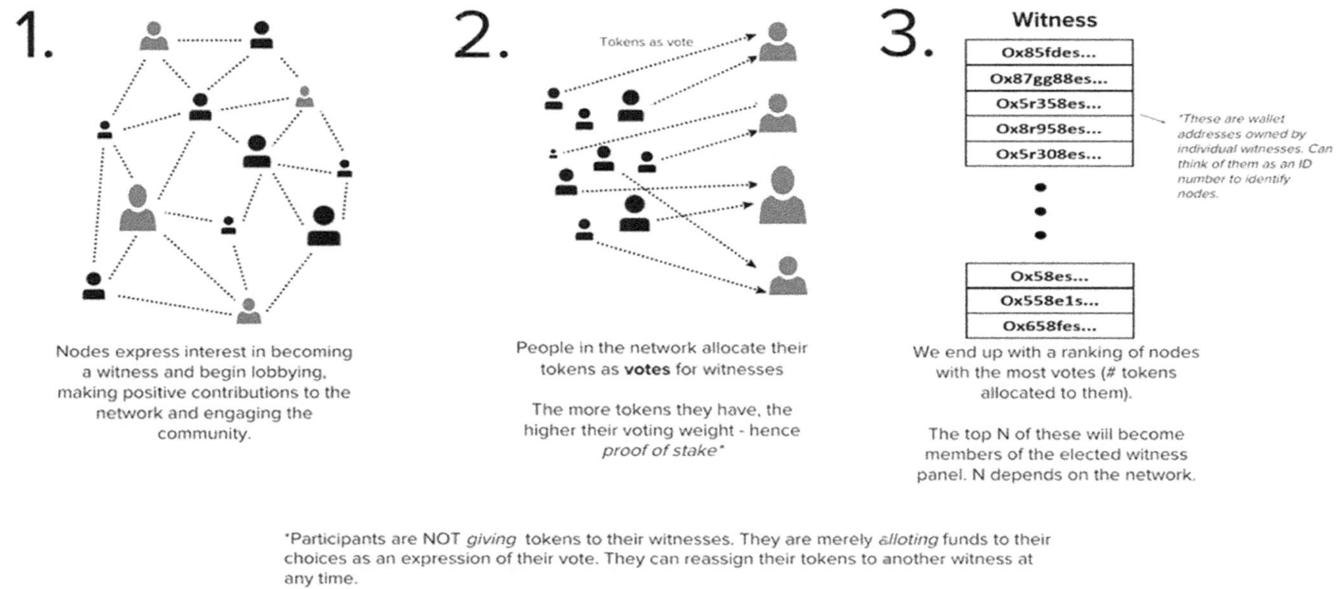

Figure 7.2 Electing witness in a DPoS network [9].

a block, the block production phase changes. Witnesses take turns according to their reputation and also depending on the number of votes they receive. Figure 7.3 depicts the scenario. If a witness misses a block—for instance, their server goes down—the task falls to the next witness in line, so block production remains fast and reliable. Witnesses are rewarded through transaction fees or sometimes block rewards, which they can opt to redistribute to their voters, invest in network development, or donate to charity. This creates an accountability for the system where stakeholders actually have a say in applying rewards.

A network usually restricts the amount of top active witnesses—for instance, on a network, this could simply be the top 100. In such competitive scenarios, a witness then has to prove that they are very valuable—for instance, by maintaining a good track record and contributing to conversations in the community. Due to inability to pay for votes or any actions, the stakeholder eventually "un-votes" them and chooses someone else because the system is constantly dynamic or competitive.

3. *Witnesses and validators*: On DPoS systems, users are voting for two groups of people where one will act as active witnesses, while another group comprises delegates, free to provide other activities involved in governing the Blockchain. They usually involve proposing parameter changes such as block size adjustment and block reward compensation to be awarded to the witness. The stakeholders would now vote if the adjustment in the proposals should be approved and implemented or rejected. Meanwhile, block validators are responsible for the full nodes—ensuring that the blocks submitted by the witnesses are abiding by the consensus set of rules. Any person can decide to run a block validator to actively help verify the network, although there is no financial incentive to do so.

Few of the beneficial factors of DPoS are that DPoS systems are prone to attacks like double-spending for very minimal chances. PoS makes the system more financially inclusive by letting users with smaller stakes take part in the decision-making process of the network. Low entry threshold allows more participants to join the consensus process, which helps in decentralizing the system further. Unlike PoW, the consensus algorithm DPoS does not require great computational power, which makes this approach more sustainable and ecologically friendly. Since the validation of transactions does not depend on heavy computing power, DPoS Blockchains can process more transactions with greater efficiency and on a larger scale. The separation of election and block production in DPoS

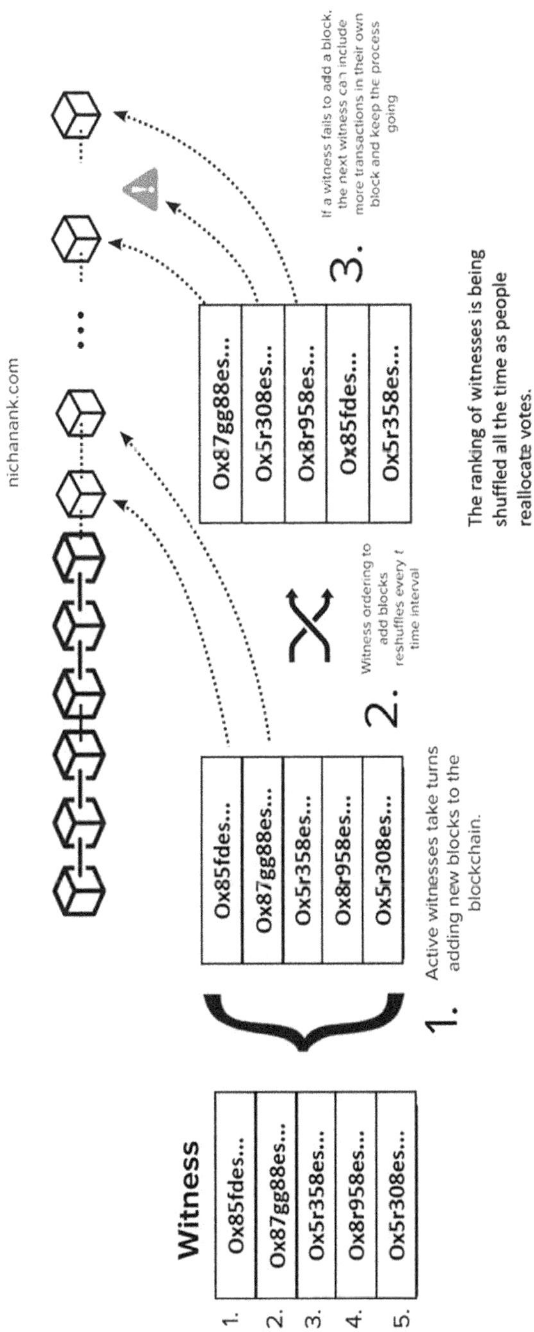

Figure 7.3 Block production in a delegated PoS network [10].

opens up avenues to experiment with various governance models for Blockchain applications.

Some limitations of DPoS are that effective network operation relies on the stakeholders being well-informed to choose honest and reliable witnesses. A small number of active witnesses can give rise to centralization if a few witnesses dominate the system [10]. Smaller stakeholders may feel that their votes are not significant and refrain from voting at all. This may bias the system toward larger stakeholders with heavier voting power.

All in all, DPoS is more democratic, efficient, and highly scalable compared to other traditional consensus mechanisms like PoS and PoW. By segregating the block production role from that of election, it allows for creativity in governance and decision-making [11]. However, a number of challenges remain in terms of decentralization and voter participation with the system.

7.4.3 Stellar Consensus Protocol

The Stellar Consensus Protocol (SCP) is a Blockchain-based system designed to provide fast, decentralized, and secure microfinance services. It was proposed by Mazieres using a method called Federated Byzantine Fault Tolerance, which makes the system resilient and robust against network faults or malicious actors [12]. SCP is open to the public, and thus anyone can participate. SCP is designed to have low-latency transactions, usually only a few seconds, akin to web transactions.

The core strength of SCP lies in its use of quorum slices, subsets of trusted nodes that help each node in reaching an agreement. In SCP, every member of the network selects a group of other members that they consider trustworthy, and these quorum slices collectively link the entire network. It lets SCP be decentralized and flexible since each node independently decides which nodes it relies on to achieve consensus.

The two major steps in the working of SCP are the nomination protocol and the ballot protocol as in Figure 7.4. The nomination of a value is where every node proposes "candidate values," which it deems must make it to the Blockchain during the nomination phase of voting. In a quorum, all the nodes propose a set of values, and each proceeds with voting for the most trusted candidate. The hope herein would be that one or several values will reach majority votes. On consensus, the system proceeds with the ballot protocol in which nodes vote for acceptance or rejection of the candidate values nominated.

Two important properties of any consensus algorithm that this system addresses include safety and liveness. Safety means every participant in the network is assured of the same result, while liveness is when the system does not get stuck and always comes up with a final decision. SCP guarantees

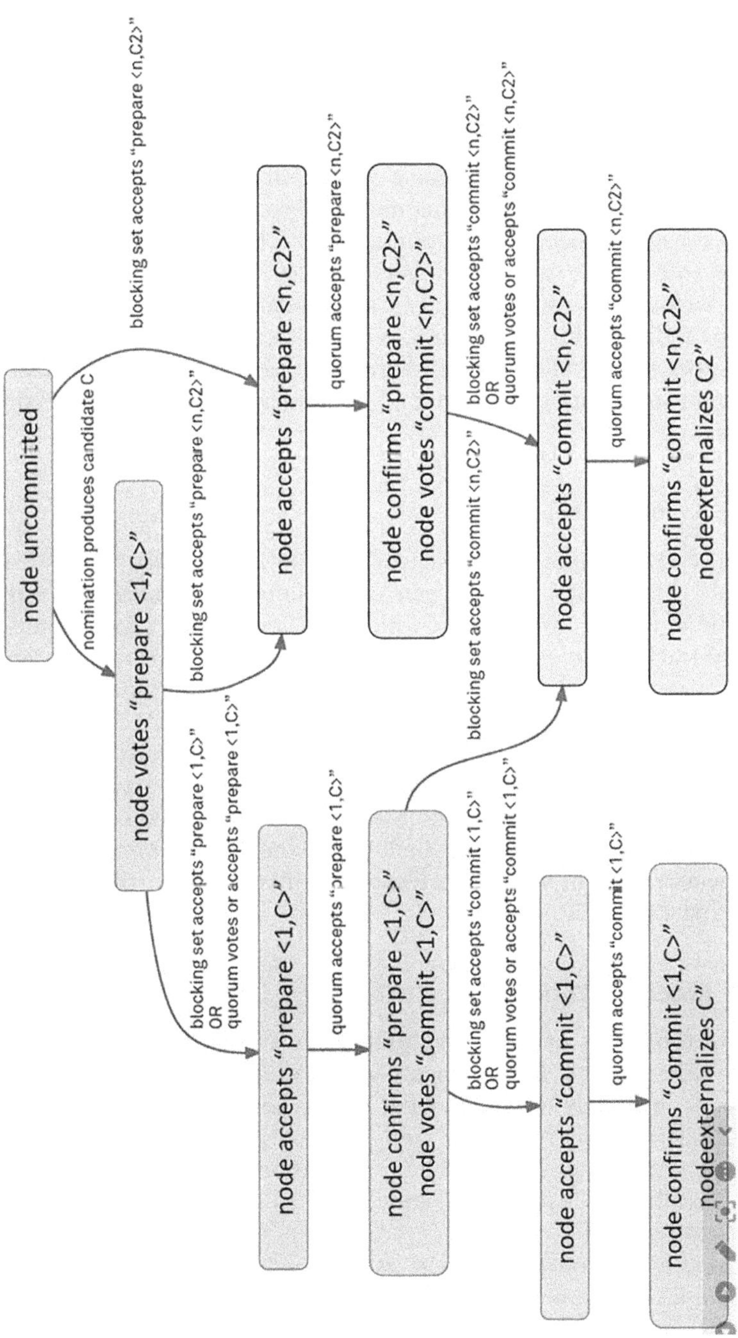

Figure 7.4 SCP balloting using federated voting [13].

both properties by using federated voting in multiple rounds to ensure that, despite possible differences in how nodes process information, the network will eventually reach a consensus.

7.4.3.1 Nomination protocol

At this stage, each node votes for a value that it trusts—for instance, "I nominate value X." These nominations are broadcast to other nodes, and as they see each other's nominations, they "echo" those votes—meaning they too vote for the values nominated by their peers. This continues until a value is confirmed, which is called a candidate. In SCP, there are many possible candidates for nomination, but an exact method narrows all these down into a single value. Even if there have been divergent node histories or different sets of quorums, nodes all do ultimately arrive at single correct decisions.

7.4.3.2 Ballot protocol

Right after the nomination protocol, the network undergoes its ballot protocol [13]. A ballot is just a pair of a counter—a number representing the round of voting—and a value, which is the nominated value. Ballots are a way of tracking the progress of the consensus process. Nodes repeatedly vote on statements about ballots indicated in Figure 7.4; as they confirm ballots, they gradually narrow down the candidate values.

Once a node reaches such a state where it may be convinced that the majority of active participants will eventually accept, it actually votes to commit a certain ballot, meaning it is assured of its value being accepted through majority vote. Once any ballot gets confirmed, messages bound by it can be acted on. In a real use, like a Stellar transaction, this may imply completing a payment.

7.4.3.3 Federated Voting

In the methodology of SCP, federation gives a chance for every member node to vote for specific statements about values and particular ballots. Each node will only vote for what it perceives, or believes in, based on its preference of outcomes, and the votes get aggregated to reach a desired consensus. Nodes echo nominations and votes from other nodes: this allows SCP to proceed with federated voting for many statements simultaneously.

The nodes vote on a statement asynchronously—that is, nodes don't operate in lockstep, and they aren't at all synchronized by time, so much as by the messages they exchange with each other. This might result in nodes drawing different conclusions at different times, but the system is so set that, given enough time, an agreement will always be reached.

- *Quorum slices*: A quorum slice is a subset of nodes that each participant trusts to reach consensus. These slices, chosen by the nodes themselves, help form the overall consensus of the network.
- *Ballot and vote*: A ballot is a pair of a counter and a value. The nodes vote on "prepare" and "commit" statements regarding these ballots, and the system guarantees that once a ballot is confirmed, the value it contains can safely be acted upon.
- *Safety and liveness*: SCP is designed to be safe—all participants agree on the outcome—and live, meaning the system is constantly in motion and never gets caught in some sort of deadlock.

By combining these mechanisms together, SCP allows a very efficient and robust consensus protocol. Its applications range from financial transactions to other systems where fast and secure consensus among a set of distributed nodes is needed. The protocol is particularly well-suited for environments that need decentralized decision-making while avoiding the energy consumption or scalability problems of other systems like Proof of Work.

7.4.4 Tendermint

Tendermint is a consensus algorithm that allows a simple and simultaneously powerful solution to reach an agreement in decentralized networks. The logic is somewhat similar to that of the PBFT algorithm: pre-vote, pre-commit, and commit are the three main steps included in this process, which allow the system to reach consensus provided there is not more than one-third of faulty or malicious nodes among those participating [14].

In Tendermint [15], the verifier nodes are part of a consensus process. The verifier nodes are randomly chosen by the network to create block proposals that should gather all the transactions on the network. It works such that time is divided into epochs, each of which is allowed to define which validator will be the round-robin proposer of the next block. Tendermint does not require a view-change mechanism, one of the salient features of PBFT; it can handle the situation in the case of timeouts or the failure of a primary node by having a special-purpose empty block, thus making the flow more even and smooth, even in situations that are problematic. It is optimized for high throughput and can easily scale up to thousands of transactions per second on a globally distributed set of nodes with latency as low as around 1 second. As a result, the performance slightly deteriorates in the occurrence of malicious attacks.

The protocol follows a simple state machine that looks like as in Figure 7.5.

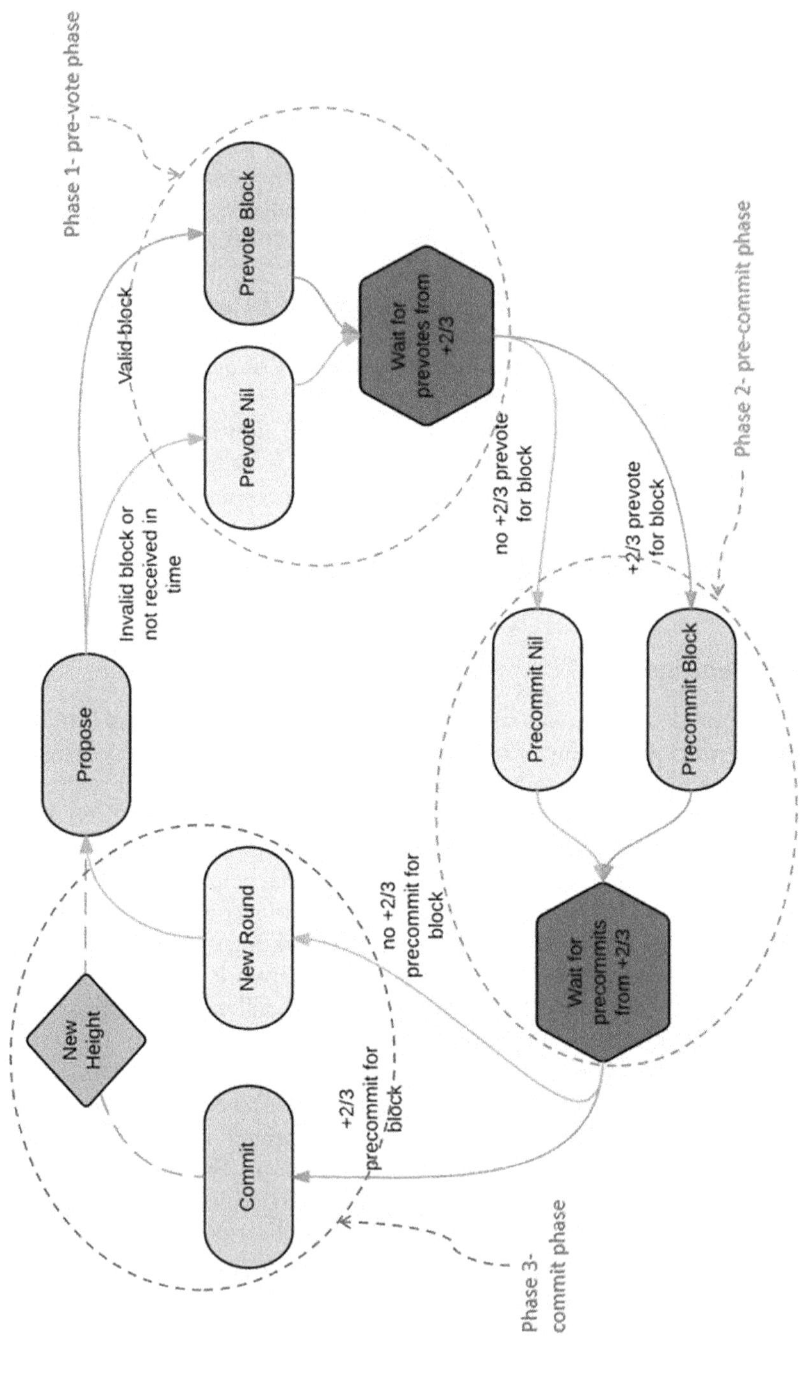

Figure 7.5 A simple state machine representation of Tendermint consensus [15].

7.4.4.1 Principal features of Tendermint

Three-step procedure: The consensus in Tendermint gets facilitated in three steps: pre-vote, pre-commit, and commit. Validators vote for prospective blocks and commit those when at least two-thirds of them have pre-committed to such a particular block.

Validators and proposers: Validators propose and vote on proposed blocks, taking turns in the proposal process. If a block fails to be committed, the system moves on to the next round, with a new validator getting an opportunity to propose.

Weakly synchronous: Tendermint rounds are based on the use of timeouts; therefore, to some extent, it represents a weakly synchronous protocol. Validators can move only after they gain sufficient information from others, with the weak possibility of delays. Despite this, it still allows blocks to get committed over time as long as at least one-third of the validators are non-corrupt.

Locking rules: Tendermint implements a set of locking rules that prevent validators from committing to conflicting blocks. More specifically, once a validator pre-commits to a block, it becomes "locked" onto that block. It can only move to another block if, during the next round, a "polka" exists on that other block. (The voting process can be imagined, in a more lively sense, as a sort of polka dance. Once more than two-thirds of validators agree on a proposal (a pre-vote), a "polka" is formed, setting the scene for the next step: pre-committing.)

Stake-based voting: Voting in Tendermint is not uniform. Each validator can have a different amount of voting power, generally determined by the amount of cryptocurrency each validator has at stake. Validators must bond their holding of the currency as collateral, which can be confiscated in case of misbehavior, creating an economic incentive to behave honestly.

In many systems where Proof-of-Stake is employed, the usage of Tendermint can be noticed because it's capable of creating decentralized applications (dApps) that are secure, scalable, and efficient. This has been designed to be easy to comprehend and implement; thus, it has become the first choice for Blockchain projects requiring high throughput and fast finality.

7.4.5 Ripple

Ripple is a distributed consensus mechanism appropriately designed to limit latency within the Blockchain [16] and also to make transactions more effective. Contrary to the majority of consensus algorithms, Ripple makes it possible for every party in developing agreement to rely on an already trusted subset of nodes which are part of the greater network. Nodes of the larger network become further divided into server and client nodes. While the server nodes are performing the consensus protocol, it is the client nodes (in the majority) used in the process of performing fund transfer. Each of these server nodes keeps a unique list of trusted nodes, considered the Unique Node List and abbreviated as UNL. The listed nodes are important regarding the consensus processes of new transaction agreements. Figure 7.6 demonstrates the process of the consensus in the Ripple algorithm.

Consensus has been achieved when 80 percent of the nodes that belong to any given UNL agree about a certain state of transaction. It repeats every few seconds on the network, which enables Ripple to quickly settle new transactions. The strength of Ripple is its potential for up to 20 percent of faulty nodes in the UNL and yet being resilient [17]. The Ripple consensus protocol can be used in many financial systems [18], especially for offering chargeback-free transactions that happen efficiently and safely.

7.4.6 Brief overview of consensus algorithms for IoT

Consensus algorithms play a critical role in the integration of Blockchain and IoT by addressing challenges like scalability, security, energy efficiency, and throughput. Table 7.1 represents the comparison of Practical Byzantine Fault Tolerance (PBFT), Delegated Proof of Stake (DPoS), Stellar Consensus Protocol (SCP), Tendermint, and Ripple Consensus Algorithm for IoT based on various parameters:

Table 7.1 summarizes the strengths and weaknesses of each consensus algorithm and presents a clear comparison that will help determine their suitability for various IoT applications.

PBFT is a leader-based consensus algorithm with high fault tolerance, capable of handling up to one-third of malicious nodes. Thus, it is highly suitable for permissioned IoT networks where devices are registered and trusted in advance. However, PBFT poorly scales for large-scale IoT networks as it suffers from high communication overhead. On the other hand, DPoS scales much better since it elects only a small group of delegates with the task of transaction validation. This reduces both latency and energy consumption, making DPoS a very good option for hierarchical IoT networks. However, DPoS brings about risks of centralization, since the selection of delegates might concentrate power in the hands of a few.

Both the SCP and the Ripple Consensus Algorithm are trust-based mechanisms. SCP uses quorum slices, enabling flexible and efficient consensus that is ideal for dynamic IoT networks such as smart cities, where devices join or leave frequently. Ripple also uses the Unique Node List (UNL) for quick, energy-efficient consensus, making it ideal for IoT applications

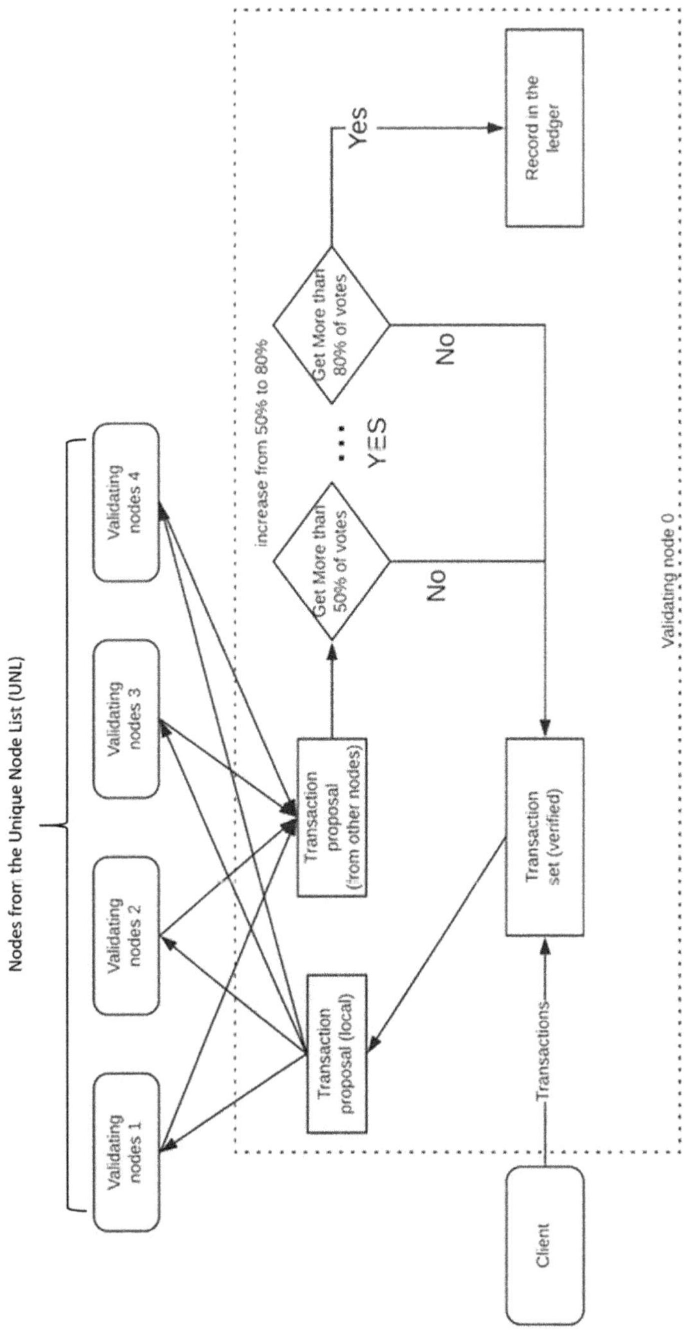

Figure 7.6 Process of the consensus in the Ripple algorithm [17].

Table 7.1 Comparison of PBFT, DPoS, SCP, Tendermint and Ripple

Parameter	*PBFT*	*DPoS*	*SCP*	*Tendermint*	*Ripple*
Consensus type	Byzantine Fault Tolerance (BFT)	Proof of Stake (PoS)	Federated consensus	Byzantine Fault Tolerance (BFT)	Federated consensus
Fault Tolerance	Handles up to one-third of malicious nodes	Depends on elected delegates; less fault tolerance	Fault tolerance based on quorum slices	Handles up to one-third of malicious nodes	Fault tolerance based on unique node lists (UNLs)
Scalability	Limited, due to high communication overhead	High, as fewer delegates participate	High, adaptable to large and dynamic networks	Moderate; suitable for smaller networks with known nodes	High; scalable for trusted environments
Energy efficiency	Low, due to extensive communication	High, as it avoids energy-intensive computations	High, with minimal computational requirements	Moderate; relatively efficient	High; energy-efficient
Transaction speed	Moderate; limited by consensus rounds	High, with fast block validation by delegates	High, with low-latency consensus	High, with immediate transaction finality	High, with near-instant confirmations
Decentralization	Low; relies on a known set of participants	Moderate; depends on delegate selection	Moderate; influenced by quorum slice configuration;	Low; validators are preselected	Low; depends on trusted UNLs
IoT suitability	Ideal for permissioned IoT networks	Suitable for hierarchical IoT systems	Best for dynamic IoT environments like smart cities	Ideal for real-time industrial IoT applications;	Best for IoT applications involving financial systems
Key strengths	High security and fault tolerance	Fast and scalable with low latency	Flexible and resilient to network changes	High throughput and real-time finality	Efficient for cross-network transactions
Key limitations	High communication overhead, less scalable	Risk of centralization due to delegate reliance	Requires trusted quorum slice setup	Scalability issues in large open networks	Requires trusted UNLs, limiting decentralization

involving financial transactions or automated payments. Nevertheless, in both configurations, properly defining trust relationships can present challenges for decentralization. Tendermint is a BFT algorithm that excels in two areas: high throughput and low latency, as it provides immediate transaction finality. This makes it ideal for industrial IoT applications that require real-time processing. It performs well in smaller, more controlled networks but struggles with scalability in open, decentralized IoT environments due to its reliance on preselected validators.

PBFT offers the highest security in small, permissioned IoT networks. DPoS and Tendermint perform well in structured environments where high performance and scalability are critical. SCP and Ripple are adaptive and energy-efficient, making them excellent options for open and dynamic IoT systems. Ultimately, the choice of an optimal algorithm depends on the specific needs of the IoT application, balancing factors such as security, scalability, decentralization, and energy efficiency.

7.5 CHALLENGES AND FUTURE DIRECTIONS

7.5.1 Current challenges

For instance, consensus algorithms like PBFT, DPoS, SCP, Tendermint, and Ripple provide some useful benefits while also facing serious challenges during application in IoT environments due to different operational requirements. Among them is the issue of scalability. A pressing concern in PBFT, for instance, is that it tends to have high communication overhead, hence impractical in IoT networks with thousands or millions of devices on a large scale. This would suggest that Tendermint works fairly well for small-sized networks, scalability issues occur once there is growth of participants involved as active validators; hence, at those levels, especially huge-scale IoT, it isn't good enough. Other concerns regard decentralization: algorithms like DPoS, SCP, and Ripple make use of trusted nodes, elected delegates, or a set of preselected validators, thus carrying the risk of centralization. This reduces trust in open IoT systems where, in fact, it is desired to avoid any concentration of control; all these algorithms are not exactly suitable for decentralized IoT applications. Energy efficiency is extremely crucial in IoT, where many devices are low-power and resource-constrained. Despite their merits, algorithms like PBFT and Tendermint can easily be power-consuming during the consensus process, hence making their feasibility a little less practical for IoT devices that shall operate on minimum power consumption.

Furthermore, latency is a very important challenge. In various IoT applications that are time-sensitive, such as autonomous vehicles or industrial automation, even the slightest delays in consensus mechanisms like SCP or PBFT could lead to operational inefficiencies and serious real-world consequences. In IoT wireless sensor networks, which generally contain various nodes and are also prone to certain attacks like a Sybil attack or some other form of Byzantine failure, fault tolerance remains very crucial. While Ripple provided a light-weight efficient solution backed by the SCP, they rely in general on carefully managed trust configurations (set of trusted validator lists as UNLs and as quorum slices) that may not adapt under dynamic conditions to the nature of any IoT network.

Finally, interoperability is a concern. Integrating into larger Blockchain platforms and allowing for the fair sharing of data between different IoT networks using these consensus algorithms is very difficult. Moreover, this leads to minimal collaboration or seamless integration among the different IoT ecosystems. The aforementioned challenges put into light the importance of more specialized, lightweight, and flexible consensus mechanisms in solving the peculiar demands of IoT systems. Solutions should handle issues such as scalability, decentralization, energy efficiency, and security while ensuring seamless integration and adaptability in dynamic environments.

7.5.2 Societal impacts of consensus algorithms for IoT

The adoption of consensus algorithms like PBFT, DPoS, SCP, Tendermint, and Ripple in IoT networks has far-reaching societal implications, reshaping industries and daily life while introducing new challenges. These algorithms help enhance data security and build trust in IoT systems, which is crucial for applications such as healthcare, where IoT devices securely transmit patient information, and supply chains, where they ensure the authenticity of products. For example, PBFT and Tendermint offer strong fault tolerance, ensuring that critical systems, like industrial automation or energy grids, remain reliable even in challenging conditions. SCP and Ripple, with their focus on energy efficiency and scalability, are well-suited for fast-paced financial transactions and smart city applications, improving everything from urban management to overall quality of life.

Despite their benefits, these technologies also raise several concerns. Centralization risks are prominent with DPoS, Ripple, and SCP, as they depend on trusted validators or delegates, which can lead to imbalances in power and limit access to decentralized IoT networks. While energy consumption is lower for DPoS and Ripple, it remains a challenge for sustainability, particularly as IoT networks expand globally. Additionally, the massive amount of data generated and stored in IoT systems can exacerbate privacy risks, raising concerns about surveillance and the potential misuse of information.

From an economic standpoint, these algorithms encourage growth by enabling new business models, such as decentralized IoT marketplaces, but they also risk creating technological dependency. As societies rely more on the efficiency and security of consensus-driven IoT systems, this dependence can become a vulnerability. To maximize the positive societal impact of these technologies, it's crucial to strike a balance between decentralization, efficiency, inclusivity, and sustainability in their deployment across IoT ecosystems.

7.5.3 Future directions and emerging trends in consensus algorithms for IoT

The future of the consensus algorithms for IoT—whether PBFT, DPoS, SCP, Tendermint, or Ripple—lies in overcoming the current limitations while addressing the specific needs of the IoT environments. Emerging is a trend in the development of lightweight, scalable algorithms to handle the huge scales of IoT networks, particularly those involving low-power devices. They work their way from PBFT/Tendermint at the local scale outwards, peering upwards with higher layers of consensus—a trend gaining widespread appreciation because of their efficiency between complete centralization and maximal decentralization. Hybrid consensus models combine some elements from each other; combining elements from DPoS and/or SCP with the latest, more fault-tolerant one can offer advantages such as Ripple using techniques like directed acyclic graph (DAG) or even Sharding to boost throughputs and tolerances for various devices and sensors within ever dynamic IoT systems.

Another promising direction is the integration of AI and ML into the consensus process: AI-driven optimizations can be used to predict malicious behavior, adapt configurations of quorum—as in SCP—or even select delegates dynamically, as done in DPoS, fortifying resilience and security. Energy efficiency in consensus protocols is turning increasingly important for the sustainability of IoT deployments, with concepts like Proof of Space or Proof of Reputation gaining momentum. Tendermint and Ripple are especially oriented to low latency and thus real-time finality, which makes the former fit particularly for IoT applications that require immediate reactions, like autonomous cars or real-time industrial control.

Another focus is interoperability—developing cross-platform consensus standards that will realize the smooth collaboration of various IoT networks and Blockchain systems. Besides that, consensus protocols are evolving to embed privacy-enhancing technologies like zero-knowledge proofs, making the transactions of data in IoT ecosystems secure but not at the cost of infringing on privacy.

These trends are a developmental process toward more adaptive, efficient, and secure consensus algorithms which could suitably meet the demanding diversified landscape of IoT. Just as these develop further along with their respective evolvement, the transformative application will also proceed across different industries and societies.

7.6 CONCLUSION

This chapter has been very aptly elaborated upon with the integration of Blockchain with IoT, wherein most importantly, the role of consensus algorithms as the backbone for a secure, efficient, and decentralized IoT ecosystem was discussed. We first addressed the basics of Blockchain, its relevance to IoT space, and how it aids in solving some very inherent problems of security, scalability, and trust issues of the integrated system. Such integration can lay the foundation for unprecedented applications in many fields. Different applications of Blockchain and IoT were also highlighted, and then their transformative power in different industries such as healthcare, supply chain, smart cities, and others was discussed. Further on, the concept of consensus was elucidated: how it evolved over time, what mechanisms secure the consistency and operational functionality in such a system. Then this chapter elaborated on some consensus algorithms like Practical Byzantine Fault Tolerance, Delegated Proof of Stake, Stellar Consensus Protocol, Tendermint, and Ripple, considering strengths, limitations, and their possible applications to certain IoT use cases.

It also outlined the challenges facing these algorithms at present, which include scalability, energy efficiency, and decentralization. Further, societal implications of the technologies were discussed, including benefits brought about by the technologies, such as enhanced security and trust, and risks, including privacy issues and tendencies toward centralization. Lightweight protocol designs, hybrid models, AI-driven optimizations, and privacy-preserving technologies will be the drivers for future consensus algorithms. Their focus is on overcoming the existing issues and better meeting the evolving needs of IoT ecosystems.

In fact, consensus algorithms form the backbone of Blockchain-enabled IoT systems, owing to the security, reliability, and scalability associated with them. If their present limitations are overcome and recent trends are welcomed, new opportunities might come along in using such algorithms for shaping the future not only of IoT but its impact on society and industry.

REFERENCES

[1] Marshall Pease, Robert Shostak, and Leslie Lamport. 1980. Reaching agreement in the presence of faults. *Journal of the ACM* 27, no. 2 (April 1980), 228–234.

[2] Satoshi Nakamoto. 2008. Bitcoin: A peer-to-peer electronic cash system.. http://bitcoin.org/bitcoin.pdf
[3] Built In, "What is a consensus mechanism in blockchain?" [Online]. Available: https://builtin.com/blockchain/consensus-mechanism. (accessed: November 23, 2024).
[4] Castro, M., and B. Liskov. 1999. Practical byzantine fault tolerance. *OsDI* 99, 173–186.
[5] Lamport, L. 2019. Time, clocks, and the ordering of events in a distributed system. In *Concurrency: The Works of Leslie Lamport,* L. Lamport, Ed.: New York, NY, USA, pp. 179–196.
[6] Zhong, Weiyu, Ce Yang, Wei Liang, Jiahong Cai, Lin Chen, Jing Liao, and Naixue Xiong. 2023. Byzantine fault-tolerant consensus algorithms: A survey. *Electronics* 12, no. 18, 3801. https://doi.org/10.3390/electronics1 2183801
[7] Buchman, E. 2016. *Tendermint: Byzantine Fault Tolerance in the Age of Blockchains*. Ph.D. Thesis, University of Guelph, Guelph, ON, Canada.
[8] N. Kesonpat, "Consensus algorithms: Proof of stake & cryptoeconomics," *NichananK Blog*, Jun. 9, 2018. [Online]. Available: www.nichanank.com/blog/2018/6/4/consensus-algorithms-pos-dpos. (accessed: November 28, 2026).
[9] Bitshares. 2016. Delegated Proof of Stake (DPOS). Available online: https://docs.bitshares.org/en/master/technology/dpos.html (accessed: November 30, 2024).
[10] Nguyen, C.T., D.T. Hoang, D.N. Nguyen, D. Niyato, H.T. Nguyen, and E. Dutkiewicz. 2019. Proof-of-stake consensus mechanisms for future blockchain networks: Fundamentals, applications and opportunities. *IEEE Access* 7, 85727–85745.
[11] Bachani, Vishal, and Aniruddha Bhattacharjya. 2023. Preferential Delegated Proof of Stake (PDPoS)—Modified DPoS with two layers towards scalability and higher TPS. *Symmetry* 15, no. 1, 4. https://doi.org/10.3390/sym15010004
[12] Mazieres, David. 2015. The Stellar consensus protocol: A federated model for internet-level consensus. *Stellar Development Foundation.*
[13] Sankar, Lakshmi Siva, M. Sindhu, and M. Sethumadhavan. 2017. Survey of consensus protocols on blockchain applications. In *2017 4th International Conference on Advanced Computing and Communication Systems (ICACCS)*, pp. 1–5. IEEE.
[14] Zheng, Zibin, Shaoan Xie, Hongning Dai, Xiangping Chen, and Huaimin Wang. 2017. An overview of blockchain technology: Architecture, consensus, and future trends. In *2017 IEEE International Congress on Big Data (BigData Congress)*, pp. 557–564. IEEE.
[15] Tendermint, "What is Tendermint?" *Tendermint Documentation*, v0.34. [Online]. Available: https://docs.tendermint.com/v0.34/introduction/what-is-tendermint.html (accessed: November. 18, 2024).
[16] Winowo Teguh, Achmad M.Y. Sulistyono, and Mochamad, Hariadi. 2020. Cryptospatial coordinate using the RPCA based on a point in polygon test for cultural heritage tourism. *Communications—Scientific Letters of the University of Zilina* 22, 211–217. 10.26552/com.C.2020.4.211-217.

[17] Schwartz, David, Noah Youngs, and Arthur Britto. 2014. The Ripple protocol consensus algorithm. *Ripple Labs Inc White Paper 5*.
[18] Chalaemwongwan, Nutthakorn, and Werasak Kurutach. 2018. State of the art and challenges facing consensus protocols on blockchain. In *2018 International Conference on Information Networking (ICOIN)*, pp. 957–962. IEEE.

Chapter 8

MANET security

A Blockchain-based secure routing algorithm

Nitesh Ghodichor and Vaishali Patil

8.1 INTRODUCTION

More powerful, portable, easier to use, and smaller than ever before, mobile devices now execute programs and offer network services. The market for mobile computer devices has expanded significantly. Mobile ad hoc networks will power the architecture of the upcoming fastest internet network (MANETs). In the overall wireless network features of the next generation (Perumal et al., 2020; Ramezan and Leung, 2018), ad hoc networking capabilities will become more important. MANET are created dynamically using an independent system of mobile nodes that interact over wireless links without the need for a central authority or network infrastructure. Random node movement, especially free node movement, arbitrary joining and leaving of the group, causes the wireless topology of the network to vary quickly and unpredictable. Such a network may function independently, or it may join the far-reaching Internet network. MANET don't require any established infrastructure; they can function independently or connect to the Internet, a larger network. Since they don't need any fixed infrastructure, like base stations as in wireless infrastructure-based systems or cellular networks, MANET are also known as networks without infrastructure. An ad hoc network is known as a multihop wireless ad hoc network because it uses many hops to route packets on established routes between nodes (Mukhedkar et al., 2020).

8.2 REVIEW OF LITERATURE

8.2.1 Attack type

The current attacks are classified into two major groups: attacks on data traffic and attacks on control traffic. Attacks on data traffic force nodes to either drop or delay the forwarding of data packets that are passing through them. While some attack techniques specifically target the packets belonging to the victims, others simply discard every packet independent of the sender

DOI: 10.1201/9781003584728-8

nodes. End-to-end delays could increase and service quality could be drastically reduced as a result. Additionally, a huge loss of crucial data results from this control traffic attacks on mobile ad-hoc networks (MANET) (Borkar and Mahajan, 2018) are inherently vulnerable due to several fundamental features. These include an open medium, distributed nodes, the autonomy of node participation (where nodes can join and leave the network at will), the absence of a centralised authority to enforce network security, and the necessity for distributed coordination and cooperation. These problems prevent MANET from utilising the present routing protocols (Abdelshafy and King, 2016). Attacks on control and data traffic have an effect on the current wireless network, and we shall see how the following attacks affect it.

It is described in the above discussion to protect mobile Ad hoc networks against such attacks, and for this reason, existing routing algorithms like Ad hoc On-Demand Distance Vector (AODV) (Dongarwar and Ghodichor, 2017) and Dynamic Source Routing (DSR) are being improved.

8.2.2 Review of secure routing and optimisation algorithm

- U. Srilakshmi et al. (2022) proposed a routing algorithm for MANETs that employs optimisation techniques to enhance routing reliability, security, and energy efficiency. Cluster heads (CHs) are selected following the application of the fuzzy clustering method, which utilises the indirect, direct, and recent trust values associated with each CH. Trust ratings were also found to be effective in valuing nodes. The cluster heads utilise multi-hop routing through the proposed protocol, which identifies optimal routes based on latency, performance, and connectivity within the defined course parameters.
- In their March 2020 article, Neenavath Veeraiah and B. T. Krishna present a method for safe multipath routing and intrusion detection in MANET. This approach employs an optimisation algorithm to facilitate effective multipath routing. To address energy and security challenges in MANET, the authors utilise cluster head (CH) selection and intrusion detection algorithms, specifically fuzzy clustering and fuzzy Naive Bayes. The selection of CH is based on indirect, direct, and recent trust values associated with each CH, with value nodes determined by trust ratings (Veeraiah et al., 2021).
- In 2020, Deepak Sinwar, Nisha Sharma, DSDV, AODV, and AOMDV examined an ant colony optimisation algorithm focused on shortest path determination in MANET. Ant-colony optimisation (ACO) and particle swarm optimisation (PSO) are two methodologies based on swarm intelligence (SI) that have been employed to determine the optimal communication path among nodes. The protocol requires optimisation to identify the most effective method

for prolonging communication availability. Numerous routing issues can be addressed through SI-based techniques. ACO provides optimal packet delivery speed, throughput, power efficiency, and minimal packet delay.

- M. S. Usha and K. C. Ravishankar (2021) propose NETAR (new energy-efficient trust aware routing) as an innovative enhancement to the traditional AODV protocol. This method integrates a trust-based security framework within MANETs, employing neighbour-node trust rate estimation, bandwidth assessment, energy predictions, and detection of malicious behaviour. The approach aims to enhance performance metrics such as packet delivery ratio (PDR).
- In January 2015, B. Madhusudhanan et al. developed a mobility-based key management approach for enhancing multicast security in mobile ad hoc networks. The anticipated stability index of a node, determined by link availability and mobility, is utilised to initially categorise them. In a multicast tree, each weak node is associated with a strong parent node, and conversely, each strong parent node is linked to a weak child node. A session key-based encryption mechanism is employed for the transmission of multicast data. The initiator node routinely performs the rekeying process. The constant rekeying interval, determined by the node type, significantly reduces the rekeying overhead associated with this technique.
- The 2018 study examines the effectiveness of the optimised dynamic source routing protocol (DSR) for mobile ad hoc networks (MANETs). The Firefly algorithm is employed to modify the traditional DSR method for identifying optimal paths between communication nodes. The proposed approach for MANET utilises the Firefly algorithm to enhance DSR routing efficiency through structured packet transfers from the source to the destination node. The optimal path is determined by considering network quality, node mobility, and end-to-end delay (Prasath and Sreemathy, 2018).
- Bhagyalakshmi proposed the Q-AODV protocol (Bhagyalakshmi and Dogra, 2018): The flood control ad-hoc on demand distance vector routing protocol, referred to as Q-AODV, seeks to minimise the total volume of control packets transmitted by network nodes by reducing the number of intermediary nodes engaged in the route discovery process. The broadcast storm of the route request (RREQ) is regulated by the queue length of the node. The queue vacancy ratio is compared at each intermediate node to the random number appended by the source to the RREQ. If the randomly generated number is smaller than the proportion of queue vacancies, the intermediate node disseminates the RREQ packet. This prolongs the network's lifespan and enhances quality of service (QoS) metrics by diminishing the quantity of overloaded nodes transmitting RREQ packets. The

proposed algorithm, Q-AODV, enhances AODV by identifying a less congested route through the assessment of queue availability. The proposed approach In comparison to AODV, QAODV somewhat decreases jitter, average end-to-end delay, and throughput.

Blockchain

Table 8.1 Survey on Blockchain and cryptography mechanism in MANET

Table 8.1 Survey on blockchain and cryptography mechanism in MANET

S. No.	*Year*	*Journal / Conference*	*Description*	*Findings*
1	2019	WSET, IJCIE	The implementation of novel processes enhances the transmission of messages and information, along with the coordination of autonomous devices (Mouchfiq et al., 2019).	A security approach known as "MPR Block chain," which is founded on Blockchain technology, is more aligned with the requirements of our group operating in ad hoc networks (Mouchfiq et al., 2019).
2	2020	CSNet Conference	This involves the requirement for a consensus system, the block graph protocol, to be resilient to network partitions (Cordova et al., 2020).	The block graph data structure is upheld, integrating a group management system that adapts to alterations in network topology and supplies relevant network topology information to the block graph framework (Cordova et al., 2020).
3	2020	COMSNETS Conference	It uses Blockchain to illustrate reputation-based routing in MANET (Careem and Dutta, 2020).	Using the Blockchain and a routing algorithm, packet delivery has improved overall (Careem and Dutta, 2020).
4	2016	ICCSITSS Conference	The Secured Hash Algorithm (SHA3-256) is presented as a method for ensuring secure routing in mobile ad hoc networks (MANETs), utilising a hybrid routing strategy (Dilli and Reddy, 2016; Khullar, 2020).	He employed the Hashed Message Authentication Code (HMAC) to ensure data integrity and authenticity. The routing method employed was the Zone Routing Protocol (ZRP) (Dilli and Reddy, 2016).

8.3 MATERIAL AND METHODS

8.3.1 Attacks in MANET

Objective: Wireless links used by MANET provide an environment that is vulnerable to threats. MANET is subject to assaults that fall under the headings of routing attacks, attacks on control and data traffic, and attacks that deny services.

In ad hoc networks, three types of dangers have been found.

- *An amateur attacker*: These attackers are experts at sniffing wireless networks or launching denial of service (DoS) attacks.
- *A professional attacker*: These attackers specialise in man-in-the-middle attacks, Sybil attacks, and layer 2 high jacking.
- Thirdly, there's the well-funded enemy, who puts a premium on rushing and wormhole attacks. In this type of attack, the perpetrator also aims to confiscate electronic gadgets.

8.3.2 Ad hoc network's vulnerabilities

Routing, the utilisation of wireless connectivity, and the auto setup methods of ad hoc networks all present vulnerabilities. All around the network, the erroneous routing information generates invalid entries in routing tables. Since routing operates on a hop-by-hop basis, intermediate nodes must cooperate. An evil node drop alters or blocks packets. This is a lack of cooperation. A node controls traffic by occupying a key geographic location. Internet Control Message Protocol (ICMP) router advertisements, neighbour solicitation messages, and Dynamic Host Configuration Protocol (DHCP) messages are susceptible to misleading answers when using an auto-configuration approach. These responses either serve as the basis for calculating IP addresses or serve as the basis for verifying that a given IP address does not actually exist. When a malicious node (Thiagarajan et al., 2021) impersonates a new node and uses the same address, duplicate address detection (DAD) happens and prevents the new node from connecting to the network. Due to their constrained computer power, ad hoc networks are vulnerable to DoS attacks. Nodes are taken prisoner physically. It can be difficult to tell what is good and negative behaviour because of several inherent features of ad hoc networks. This includes packet loss owing to a weak wireless link and the absence of nodes due to movement and power limitations.

8.3.3 Security in MANET

Numerous applications are now possible because of mobile nodes' collaborative processes. Information availability, integrity, and secrecy are all aspects of traditional security. The information owner establishes a security policy and puts that policy into action by taking appropriate precautions. For the open network, policies cannot be defined (mobile ad hoc networks). The effectiveness of a partnership is largely dependent on the moral character and integrity of its participants as expressed through trust and reputation-based processes (Jaydip Sen, 2010; Veeraiah et al., 2021).

8.3.4 Current defence mechanisms in MANET

A range of secure routing mechanisms have been developed and implemented to ensure network security in MANET. This study examines various secure routing mechanisms, including key-based routing, intrusion detection systems, and watchdog processes.

8.3.5 Need of trust-based system

Key decision-making tools MANET has trust and reputation management schemes. That makes it easy to distinguish which MANET nodes should receive data or services from others and which should send packets. This decision-making procedure protects MANETs from internal attacks when cryptographic security fails. A mobile ad hoc network of nodes for data routing and sharing has been useful. Setting up the network is simple. Decision-making is easier when people trust one other (Zach and ChainLinkGod, 2022). Trust makes packet routing, file sharing, and file download service providers easier to choose. Dropping packets, delaying packet forwarding, withholding files, or refusing data damages node trust. Even the most trustworthy node can behave this way during a key activity. Then, the trust computation engine de-trusts all nodes and declares the most trustworthy malignant. If so, communication will occur via less reliable nodes or not at all. Since we integrate node state in trust computation, this problem and wasteful trust value decreasing no longer exist. Trust computation systems often produce false-positives and -negatives, resulting in untrustworthy results. Nodes generally contest accusations of unfairly complementing or disparaging others.

There is no central authority that controls MANET. It is not possible to forward packets from one node to another using pre-generated fixed secure routes since the network is quite dynamic. A node may need additional nodes for packet routing. On an open network, the data owner cannot regulate security. Cryptographic security fails to safeguard communications and authenticate nodes (Awan et al., 2022). They cannot grade service quality or survive packet dropping, delayed delivery, or haste. Reputation and trust systems are helping customers choose routing nodes and buy services.

8.3.6 Definition of trust

Many trust models for safe node communication based on group recommendations have been suggested to promote computational network collaboration. A person's trust in any participating node's behaviour. Trust management provided and managed security relationships and policies differently than other security services. Trust management in MANET determines information and node belief levels, identifies intrusions, and provides security services including key management (El-Bashary and Abdelhafez, 2015), authentication, access control, and node revocation. Therefore, specialised computational methods should be utilised often to determine confidence. Trust in MANET must be built by quickly identifying and isolating intruders to prevent them from exploiting network resources. One can misjudge trust based on self-detection of inappropriate conduct. Selfish nodes can't be detected by a node that's not sending packets. Cooperation between adjacent nodes is required (Jothi et al., 2016, Jothi, J. Sathiya. 2018). Trust management is needed in MANET when nodes send data without knowing them or wanting to construct a trusted network (Pilankar, Pooja, and Puja Padiya, 2016). Cryptography and intrusion detection systems (IDS) are used to prevent security breaches; however, malevolent attackers often refuse to comply and compromise network authenticity and secrecy, making it hard to identify these nodes. To solve these problems, trust is needed. Trust is founded on recommend ability. We decide whether to communicate with a node depending on its dependence on others. Another security method may not be able to fix MANET security (Singh and Singh, 2016) vulnerabilities due to a problem. Trust management is needed because nodes' configured behaviour can block dangerous communications and allow trustworthy nodes to communicate freely (Awan et al., 2022).

Direct forwarding proof: Signed, praised information: The calculation of a direct trust is done on another node. Discussion with another node forms its foundation. This divides the data into positive and negative events. Positives include route request, reply, error, and data flow. Faulty behaviour includes flooding, packet loss, changing the route sequence number, adding or removing routes, and faking. Since MANET (Awan et al., 2022) continuously calculates trust, T values are limited to –1 to +1.

recommended trust or indirect evidence: Nodes exchange parked vehicles forwarding mechanism (PFM) counts when contacted: The SMRTI recommender gives advice about another node. Recomandees are nodes that have been recommended. Here, the routing packet provides recommendations. The node's readiness to transfer more data or discard the packet is determined by Secure MANET routing Trust Intrigue (SMRTI). SMRTI only makes communication recommendations once, thus malicious routes cannot change them. For instance, Node B recommends Node A to Node C. All trustworthy network nodes will do this and save the T value of the previous node in the data flow sequence. When computing suggested trust, positive and negative events are considered to prevent rogue nodes (Awan et al., 2022). Building trust is hard. It is vital for ad hoc network security routing. MANET and multi-hop routes lack infrastructure; hence a trusted path is needed to transport packets even with hostile nodes. Packet loss and retransmission in MANETs are unpleasant. A trust model creates trusted channels, increasing packet delivery. Trusted paths ensure nodes receive requested services.

8.3.7 Scope of trust

Parties must choose to trust. Trust is enabled and referred by dependent party.

Functional trust: Reliable party finishes task. Values are based on trust. Success requires task completion.

The trustworthy party advises a party that executes the function. Direct experience builds trust. The suggestions of others create indirect trust.

Trust properties: These attributes are measured by trust. Trust matters in a certain environment. The function of uncertainty: The entity can finish its tasks.

Probabilistic trust: Continuous or discontinuous numbers may signify trust. A believes in B, although B may not. Trust is not symmetrical or identical in both directions.

Transitive property builds trust. If A Trusts B and B Trusts C, A trusts C. While they may differ, A Trusts B and B Trusts C do not indicate A Trusts C. Trust is subjective and not transferable mathematically. Trustworthiness is disputed by two parties.

8.3.8 Problem statement

This work implements a trust system based on Blockchain technology to identify malicious nodes, standardise trust computation through a hashing (Ethereum, 2021) function to prevent packet dropping, determine the shortest path between source and destination nodes, and ensure packet delivery is protected from wormhole and black hole attacks (Patil et al., 2014).

About the title and requirement to work on it: A proposed plan MANETs use secure routing protocols to prevent attacks. To secure network communication, an implementation module uses secure route discovery from source to destination and anonymous key establishment between two nodes.

It prevents malicious nodes from joining and becoming malicious, creating a safe transmission network.

Establishment of an anonymous key: Each packet transmission node creates a session key with its neighbour. Transmission of a shared secret key begins after both parties verify it. Secure route finding: Route discovery is anonymous. The source node establishes a secure path to the destination node using the shared secret session key (GeeksforGeeks, 2021).

The proposed solution ensures that external nodes are unable to access the network without first performing the necessary calculations to derive the shared secret key. Additionally, if a node turns hostile, it will be incapable of computing the data transfer secret key. This technique protects PDR, throughput, and privacy. A trusted path for routing in mobile ad hoc networks must be created to boost throughput, assert non-repudiation and lose packets, protect against various assaults, and account for node fixed and random mobility. Trust (V. S. and M. S. K., 2018) calculation methods vary, creating research obstacles and gaps.

Tracking prior behaviour and documenting all node behaviour is a difficult space. Time-based weighting increased the computational cost for each observation. Ad hoc network nodes are always energy sensitive. What stochastic characteristics should be considered for estimating lifespan and trust path? Nodes with priority work may not participate in routing. In that instance, binary judgement may not work. Unjustified praise or criticism shouldn't affect trust path calculation. Nodes deny expressing favourable or negative ideas consistently.

Other trust computation engines include fuzzy models and flow models, which were recently created by a separate author. We propose a Blockchain-based trust model. Node behaviour is uncertain. Most previous work uses probabilistic logic and ignores excellent node behaviour uncertainty. In contrast to unnecessary trust updates for a good node when a node acts maliciously during priority work, our technique examines both fixed and random mobility and the node's state before judging trust. This study gives a convincing application case.

We present an arbitrary logic approach to trust calculation. The trust and reputation computation engines of summation utilised a belief model based on subjective logic operators to calculate confidence in the suggested trust model because node behaviour in ad hoc networks is inherently uncertain. Our trust calculation includes packet forwarding and dropping between source and destination nodes. First-hand knowledge and node references can also build trust.

Attacks considered for trust system: The attack dataset listed below is taken into consideration: These attacks have an impact on MANET communication as well as the growth of schemes dependent on reputation and trust Black hole, Grey, Warm, Altered/Replayed, Sybil, Spoof Routing, Sinkhole, Hello Flood, and DOS-Apache2 are among the many attack types included in the dataset. A total of 10,028 datasets will be produced using this method. Even though the attack has been effectively identified, 3,079 of the 6,893 datasets are incorrectly classified. As many as 526 datasets show that attacks and typical behaviour are unpredictable. In summary, 9,502 datasets successfully use the confusion matrix of categorisation to categories attacks as either malicious or normal.

8.3.9 Proposed method

Secure routing algorithm with blockchain (*SRABC*): The assault dataset that follows is taken into account. These attacks impact the growth of schemes dependent on reputation and trust as well as communication within MANETs. Attack types included in the dataset include Black hole, Grey, Warm, Altered/Replayed, Sybil, Spoof Routing, Sinkhole, Hello Flood, and DOS-Apache2 (Ghodichor et al., 2022), this method will produce 10,028 total datasets.

Algorithm: The goal of the SRABC technique is to minimise the MANET delay value.

Input: Node parameter (Maximum Node 100, Parameter 5)
Output: Reduce the node parameter function(delay)
Step I: The parameters and nodes that make up the network are described in isolation.
Step II: Nodes Sr and Dt configured as sources and destinations, respectively.
Step III: Utilize the AODV protocol to establish routes between the source and destination. The route is specified as Sr, R1, R2, R3... Rn, Dt.
Step IV: For a = 1 to n, execute steps IV to VIII repeatedly.
Step V: RList find adjacent (a)
For b-1 to length (RList)
{
Step VI: Parameter = Packet Delay (b)
Step VII: Using Fuzzy Logic for parameter fuzzification
Step VIII: If the parameter Delay =low
{
Set Rlist(a) as the next communicating node
}
}
Step IX: Go to the following next node.
Step X: Finish

Employing an advanced algorithm that effectively lowers latency for numerous interconnected Blockchain nodes (Geroni, 2023; Aswal, 2024). Blockchain technology (Iredale, 2023) has proposed the creation of an innovative relinquished blockchain-based (Wessling and Gruhn, 2018) integrity system to efficiently mitigate hazards. Subsequently, the reliability of the network system is assessed.

8.4 RESULT AND DISCUSSION

In order to find a path to a location or data gateway inside heterogeneous wireless networks, the BCR (Ramezan and Leung, 2018) protocol makes use of smart contracts. Any device serving as an intermediary can ensure a path from a source device to a gateway or destination device. In a network of devices, we contrast the performance of BCR and the ad-hoc on-demand distance vector (AODV) routing protocol. We employ public ledger solutions, such Blockchain, to use for mobile ad hoc network due to the absence of confidence in a central management system and the requirement for a publicly verifiable reputation system.

The evaluation of BCR's simulation employs metrics including PDR, average throughput, end-to-end delay (indicating dropped packets), routing overhead (latency in route acquisition), block height (related to deposit addresses), minimum difficulty (number of nodes), movement speed (transmission range), and simulation time (latency in block generation), among others. Upon expiration of a route request, this section includes blacklisted addresses, a selected route, a timer for the blacklist, and the validity of the route offer.

PDR: Delivering all of the data packets to the destination is beneficial, and the PDR value increases the network's effectiveness. The PDR is the ratio of the total number of packets delivered from the source node to the network destination node to the total number of packets transmitted.

$$PDR = \left(\frac{\sum Received\,Packets}{\sum Sent\,Packets} \right) \times 100$$

Average throughput: A predetermined number of data packets are sent to the recipient in a predetermined period of time. Usually, bits per second are used to calculate it.

$$Throughput = \frac{\sum Number\,of\,Packets_{received}}{\sum TotalTime}$$

Total time required: The end-to-end or one-way delay (OWD) is the amount of time it takes for a packet to go from its origin to its final destination across a network. The widely used in-network surveillance method, in contrast to the more complex round-trip time RTT, measures merely the one-way flow from the source to the destination.

$$d_{end\,to\,end} = N\frac{L}{R}$$

Routing overhead: The amount of packets sent for maintenance and path discovery can be considered.

$$Routing\ Overhead = \frac{Sum\,of\ routing\,packets}{Sum\,of\ data\,packets}$$

Route acquisition latency: This period encompasses the interval from when a source transmits a request or discovery packet to a destination for path determination until the initial response is received. The first transmission interval is used to measure latency when a path demand must be retransmitted due to timing issues.

Block height: The block height of a specific block indicates the number of preceding blocks in the Blockchain. The genesis block represents the initial block within a Blockchain structure. Since there are no prior blocks, the Blockchain's block height is 0. Using a Blockchain length less than one is a popular method of calculating the most recent block's cumulative Blockchain height. The biggest block is on the Blockchain.

$$Block\,height = block\,chain - 1$$

Deposit address: Deposit addresses are temporary addresses that link Gate Hub deposits to user wallets on a native cryptocurrency network. For cryptocurrencies other than XRP, Gate Hub does not include the user in its register addresses.

Maximum difficulty: A quantity called difficulty controls how long it takes miners to add fresh blocks of transactions to the Blockchain. The difficulty_1_target variable can change depending on how difficulty is measured.

$$Maximum\ difficulty = \frac{max\,imum - target}{1} = \infty$$

Number of nodes: A complete node, such as a computer, contains the complete history of Blockchain transactions. The Blockchain is composed of data blocks.

Block generation latency: A common name for this is "block time." How long it takes for the network to process a request and then build the next block of transactions is known as network latency. To rephrase, it's the time it takes for a user to see their transaction recorded on the Blockchain after clicking the send button.

8.4.1 Simulation outputs

The results of the simulation are analysed in the figures that follow. The first step is to set up the Blockchain network in an arbitrary way. The next step is to set up the Blockchain inside the MANET, which will allow communication between nodes by assigning a weight to each node. By finding the shortest communication path, the MANET transaction was successfully completed. The training dataset takes into account a wide variety of techniques, and many of those attacks have been tested and confirmed in the dataset.

In Figure 8.1, we can see how to configure a MANET Blockchain node such that it has the appropriate weight. Each node in a Blockchain is given

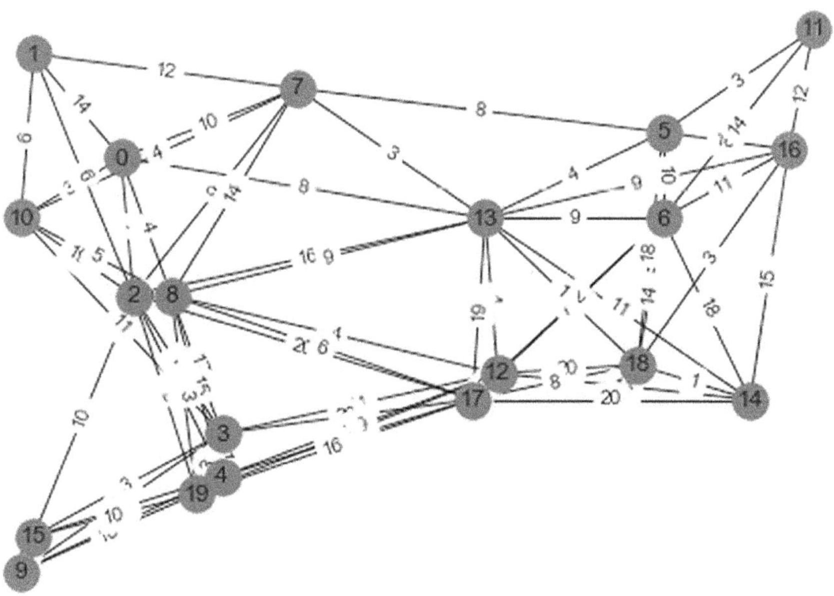

Figure 8.1 Node creation in blockchain.

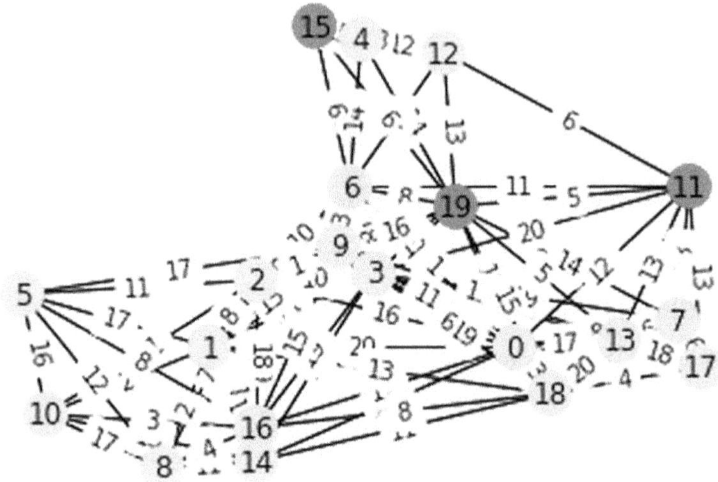

Figure 8.2 Shortest path calculation.

the appropriate weight when data is hashed and a block is generated. Then, a node is constructed using the Blockchain.

Figure 8.2 shows the procedure for choosing the fastest path to transfer data from one node to another. Eleven is the starting point, while 15 is the ending point. Hence, the shortest route is the one that starts at node 11, travels to node 19, and finally ends at node 15. If we start at node 12, it will take two nodes to get there.

8.4.2 Simulation output without Blockchain in MANET

The throughput is graphed against time in Figure 8.3, absent Blockchain security. The study reveals that the throughput is 400 packets at 20 seconds, 700 packets at 40 seconds, 950 packets at 60 seconds, 1,590 packets at 80 seconds, and 1,780 packets at 100 seconds. Figure 8.4 illustrates the average end latency over time in the absence of Blockchain. This graph illustrates the correlation between the average end delay and the duration of the pause over time. Upon reaching a pause duration of 100, the mean terminal delay is 22.0 seconds.

The ultimate state of the delay with respect to Blockchain nodes is shown in Figure 8.5. Without Blockchain, the end-to-end latency is close to one millisecond (ms), and it increases when the number of nodes reaches five.

Figure 8.6 shows the delivery ratio of no Blockchain packets over time. The PDR increases to 89.5 percent for 100s as time passes effectively. PDR

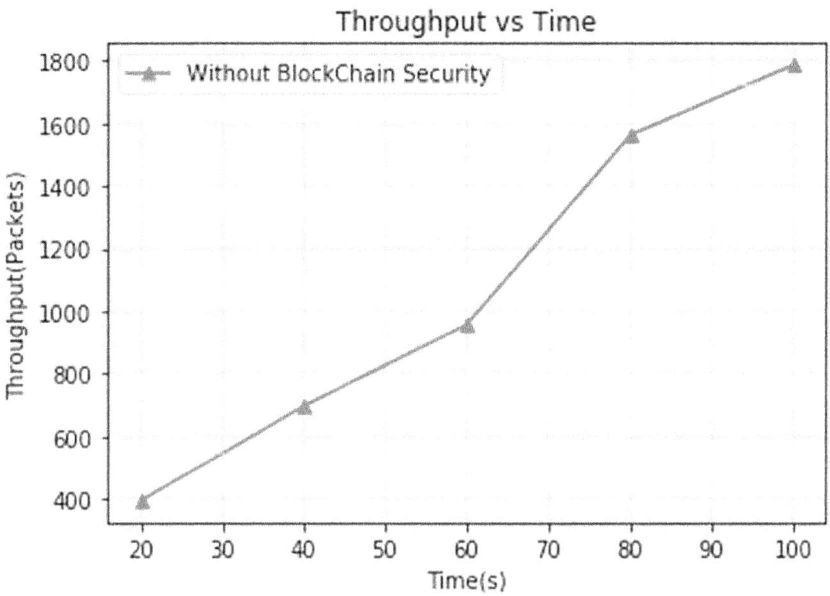

Figure 8.3 Without blockchain throughput vs time.

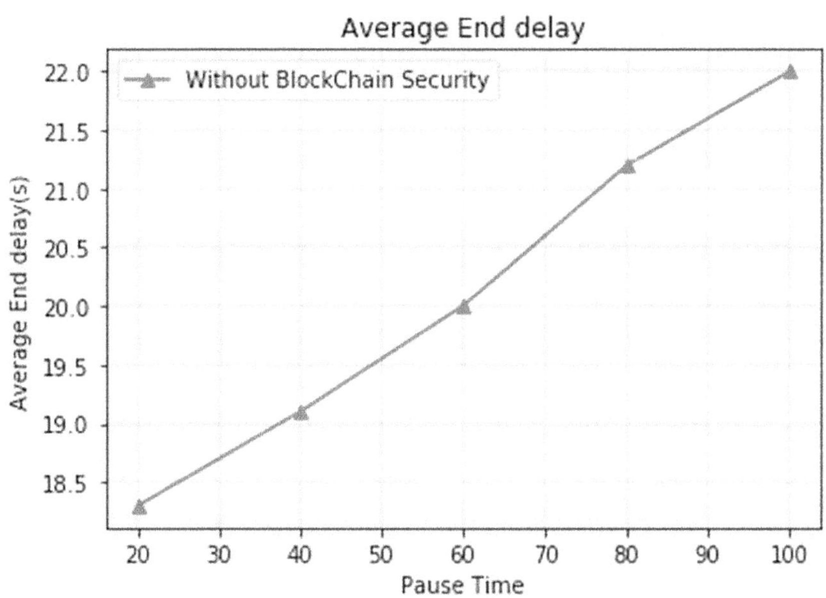

Figure 8.4 Without blockchain average end delay vs time.

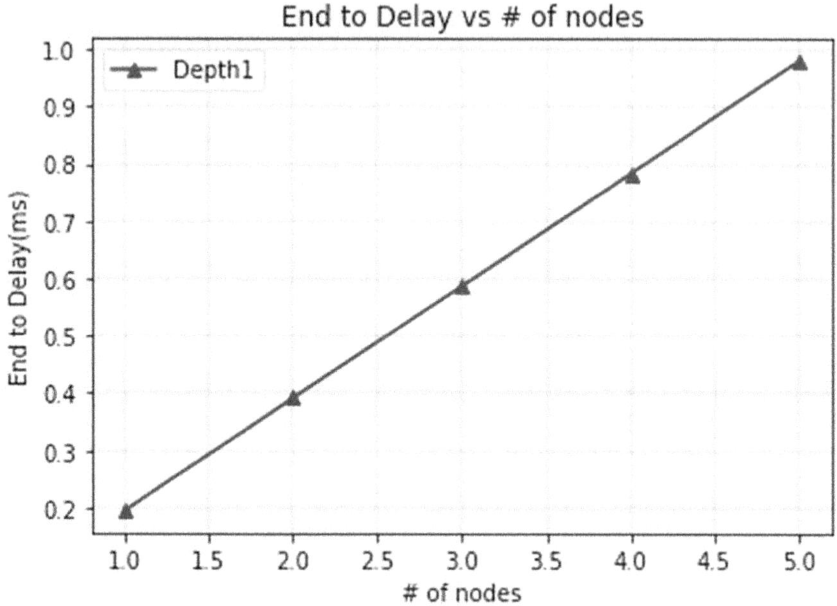

Figure 8.5 Without blockchain end to delay vs nodes.

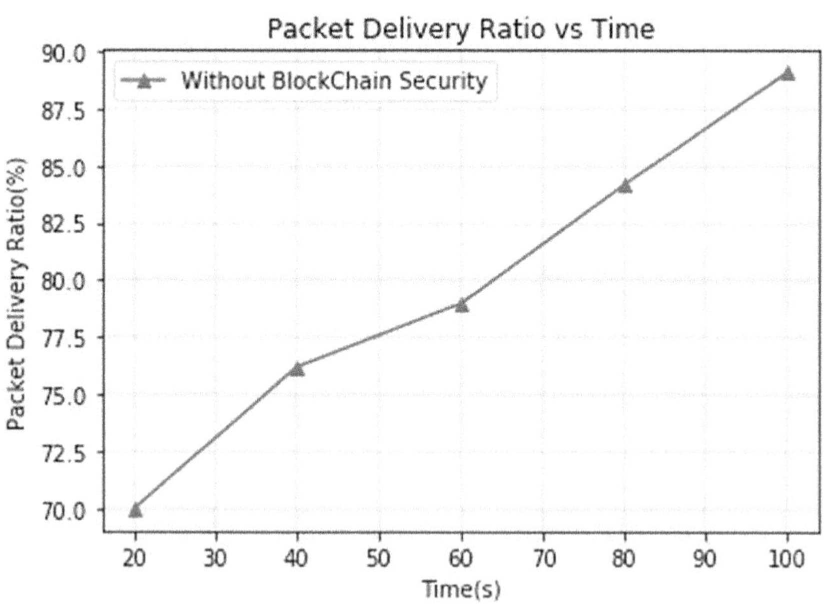

Figure 8.6 Without blockchain PDR vs time.

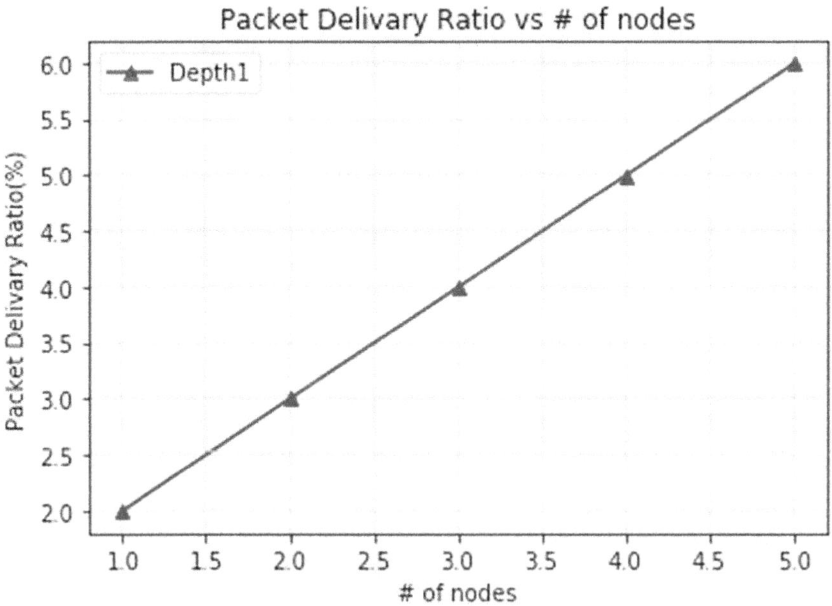

Figure 8.7 Without blockchain PDR vs node.

vs node without blockchain is displayed in Figure 8.7. In the absence of Blockchain, the PDR hits 60 percent and the node reaches 5.

The routing time versus overhead in the absence of Blockchain is shown in Figure 8.8. This graph shows that without Blockchain, the latency rises to 100s and the routing overhead approaches 69 percent.

In the absence of a Blockchain, Figure 8.9 displays the routing overhead per node. When there are more than five nodes without a Blockchain, the routing cost rises to 45 percent. The graph illustrates this.

Table 8.2 presents the parameters of MANET in the absence of Blockchain over time. The variables in the table, including throughput, average end delay, PDR, and routing overhead, were analysed for duration ranging from 20 to 100 seconds. A total of 400 packets were delivered, with an average delay of 18.3 seconds. The PDR stands at 70 percent, while the routing overhead is recorded at 10 percent over a period of 20 seconds. At 40 seconds, the throughput surpasses 700 packets, the average latency is 19.1 seconds, the PDR stands at 76 percent, and the routing overhead is 23 percent. A total of 950 packets were transmitted, resulting in an average delay of 20 seconds. The PDR achieved was 79.5 percent, and within a time frame of 60 seconds, the routing overhead was calculated to be 38.4 percent. At the 80s mark, throughput has achieved 1,590 packets, average delay stands

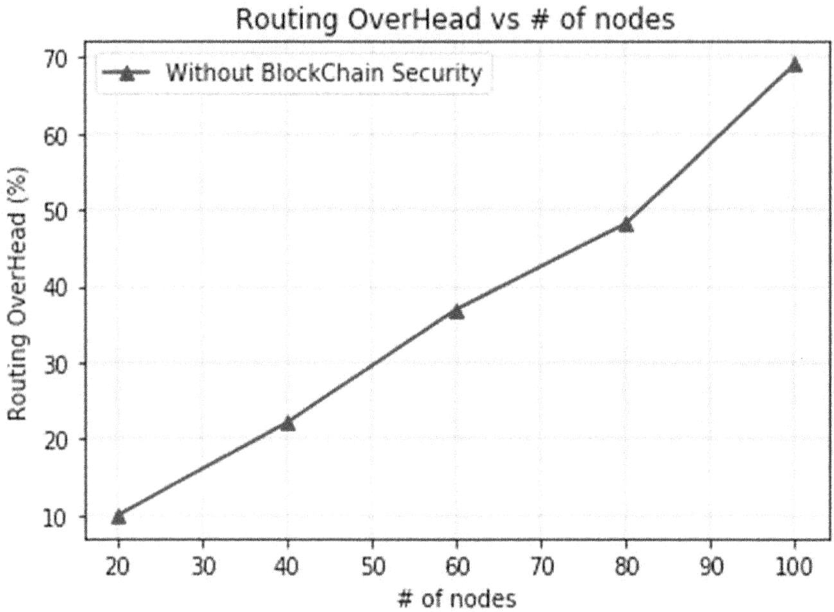

Figure 8.8 Without blockchain routing overhead ratio vs time.

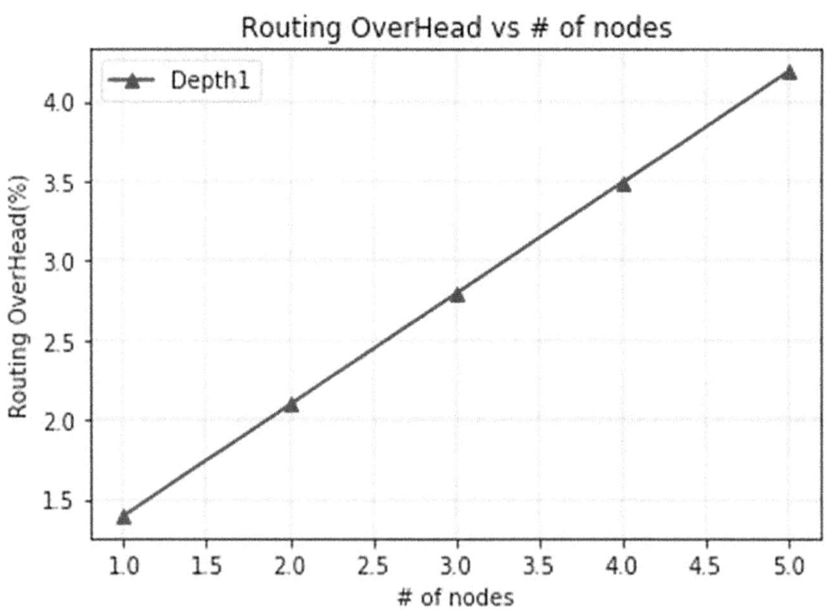

Figure 8.9 Without blockchain routing overhead ratio versus nodes.

Table 8.2 Parameters vs time for without blockchain in MANET

Time (s)	*Throughput (packets)*	*PDR (%)*	*Average end delay(s)*	*Routing overhead (%)*
20	400	70	18.3	10
40	700	76	19.1	23
60	950	79.5	20	38.4
80	1590	84	21.3	49
100	1780	89.5	22	69.8

Table 8.3 Parameters vs node for without blockchain

Node	*End to delay*	*Routing overhead*	*PDR*
1	0.2	14.6	20
2	0.38	21	30
3	0.59	28.8	40
4	0.8	35	50
5	0.96	43.6	60

at 21.3 seconds, PDR is at 84 percent, and routing overhead is recorded at 49 percent. At 100 seconds, the throughput is 1,780 packets, the average latency is 22 seconds, the PDR is 89.5 percent, and the routing overhead is 69.8 percent. Table 8.3 presents the parameters associated with nodes in a MANET that lack Blockchain integration.

Table 8.3 presents a comparison of settings and nodes for MANET in the absence of Blockchain. In addition to the five nodes, other variables were evaluated, such as average end delay, routing overhead, and PDR. The end-to-end delay measures 0.2 ms, the PDR is 20 percent, and the routing overhead is 14.6 percent for a single node. The end-to-end delay measures 0.38 milliseconds, the packet delivery rate stands at 30 percent, and the routing cost increases by 21 percent with the inclusion of two nodes. The achieved metrics include a PDR of 40 percent, a routing overhead of 28.8 percent, and an end-to-end delay of 0.59 milliseconds, utilising three nodes. The system with four nodes exhibits an end-to-end delay of 0.8 ms, a routing cost of 35 percent, and a packet delivery rate of 50 percent. Increasing the number of nodes to five leads to a reduction in end-to-end delay to 0.96 ms, an enhancement in PDR to 60 percent, and a rise in routing overhead to 43.6 percent.

8.4.3 Simulation output with Blockchain in MANET

Relevance and weight are related, as seen in Figure 8.10. The weight will gradually rise in accordance with relevance before falling. In Figure 8.11,

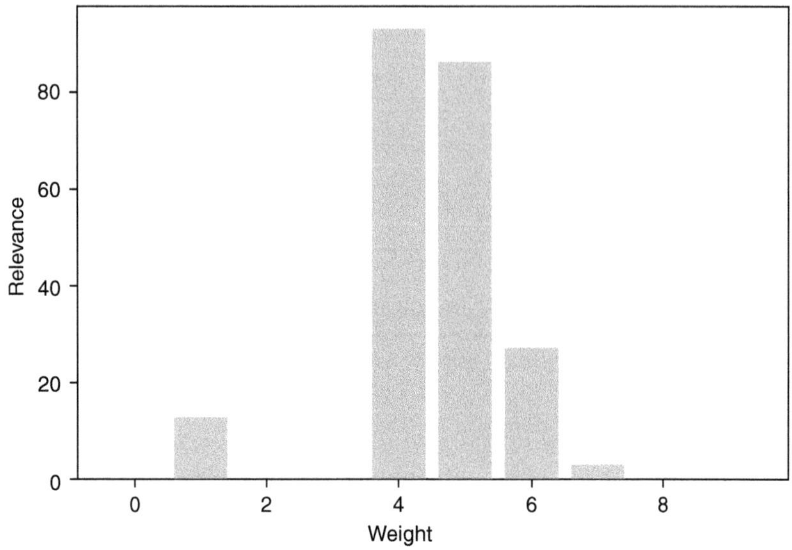

Figure 8.10 Relevance vs weight.

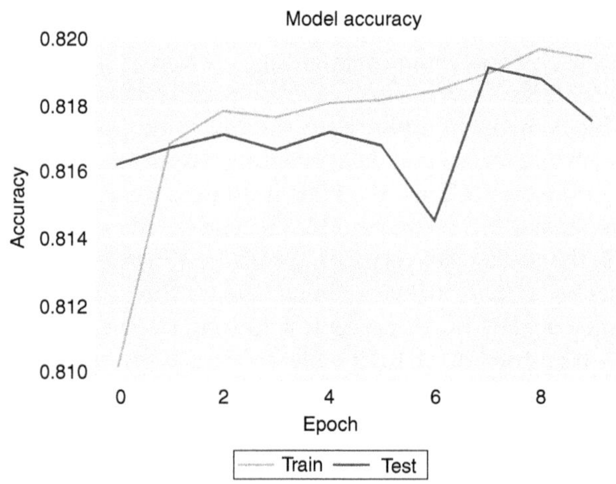

Figure 8.11 Model accuracy of the dataset.

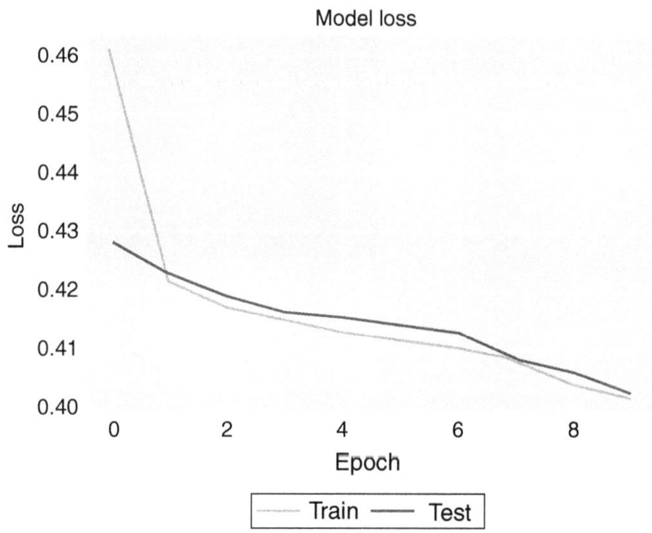

Figure 8.12 Model loss of dataset.

the data set shows that the model is correct. Plotting of the accuracy and epoch functions is done using both the train and test datasets. The accuracy of the training dataset will be substantially higher than that of the test dataset.

Figure 8.12 shows the relationship between time and loss. The loss in comparison to the test dataset will be minimal as the training dataset gets smaller. The categorisation confusion matrix is shown in Figure 8.13. Blackhole, Grey, Warm, Sybil, Altered/Replayed, Spoof Routing, Sinkhole, Hello Flood, and DOS-Apache2 attacks are among the several attack types included in the dataset (Sandeep None, 2012). With this approach, 10,028 entries will make up the final dataset. Thirty-nine datasets were misclassified, whereas 6,893 datasets successfully detected the attack. And 526 datasets are insufficient to anticipate attacks or typical behaviour. In the analysis, 9,502 datasets in all have been effectively divided into two categories: harmful and normal attacks.

Figure 8.14 depicts the correlation between the number of nodes in a Blockchain and the level of security it provides. The graph indicates that the security level will attain a maximum of 99 percent when the number of nodes is comparatively limited. As the quantity of nodes increases, the security level consistently declines until it reaches 97.5 percent.

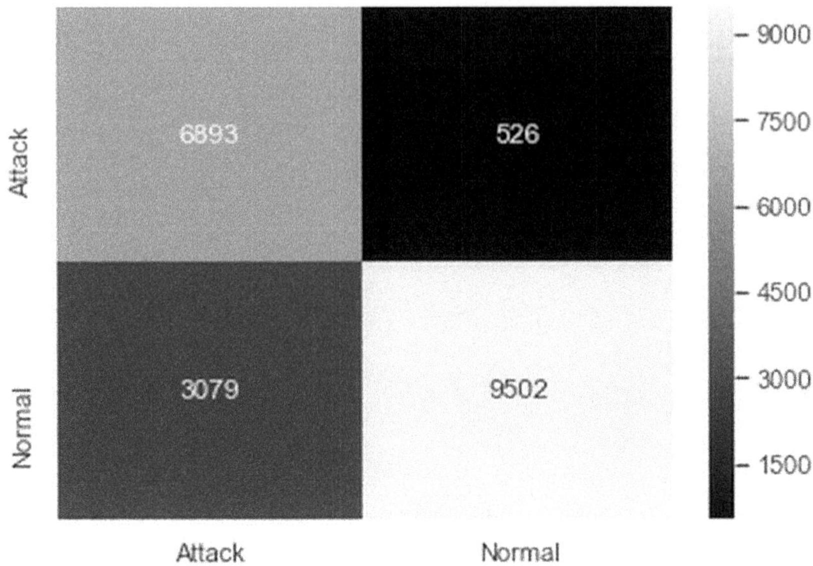

Figure 8.13 Confusion matrix.

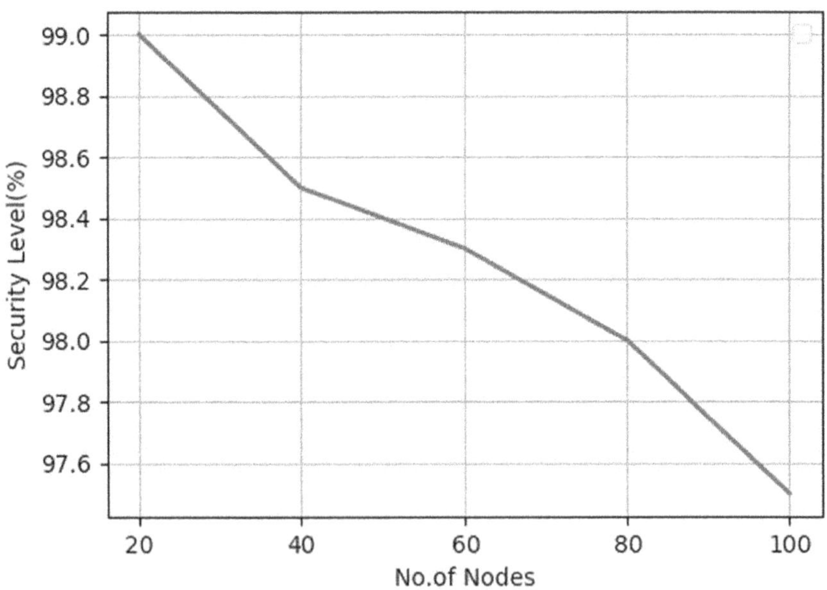

Figure 8.14 Security level in blockchain.

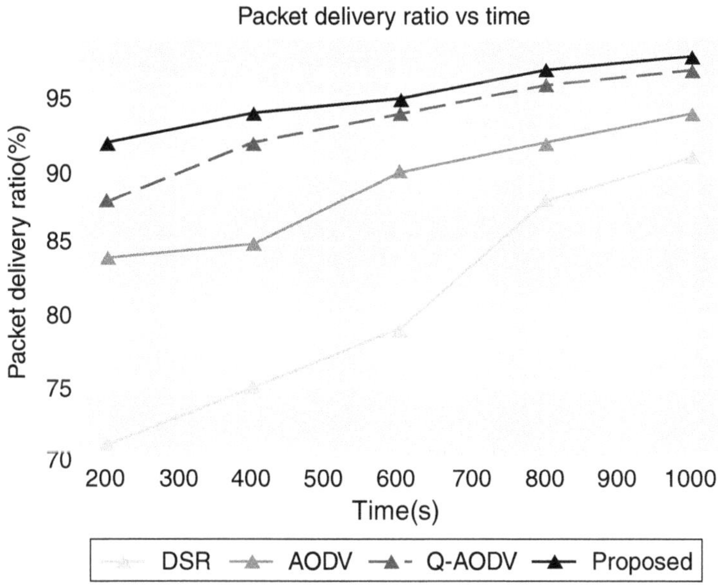

Figure 8.15 PDR comparison.

8.4.4 Comparison strategies

This section compares the performance of the proposed method with various algorithmic schemes, including AODV, QAODV, DSR, Highly Efficient Dual Authenticated Routing (HEDAR), and secured encryption technique with optimum route discovery (Borkar and Mahajan, 2016; Mouchfiq et al., 2019; Cordova et al., 2020; Careem and Dutta, 2020). Figure 8.15 depicts the operational mechanics of the comparison process.

Figure 8.15 presents a comparative analysis of the PDR in relation to prior methodologies, including DSR, AODV, and QAODV. The new strategy demonstrates a significantly improved PDR when compared to the previous methods, as illustrated in the graphic. This method effectively reduces packet loss and has been successfully implemented.

Figure 8.16 presents a comparative diagram depicting the relationship between the number of nodes and the PDR, highlighting the performance of the proposed method in relation to previous technologies, specifically Secured Encryption Technique with Optimal Route Discovery (SETORD) and HEDAR. The chart clearly indicates that the PDR has achieved its highest level when compared to the previous method.

Figure 8.17 presents a comparison of the prior strategy and the proposed strategy, focusing on packet throughput results. The proposed method

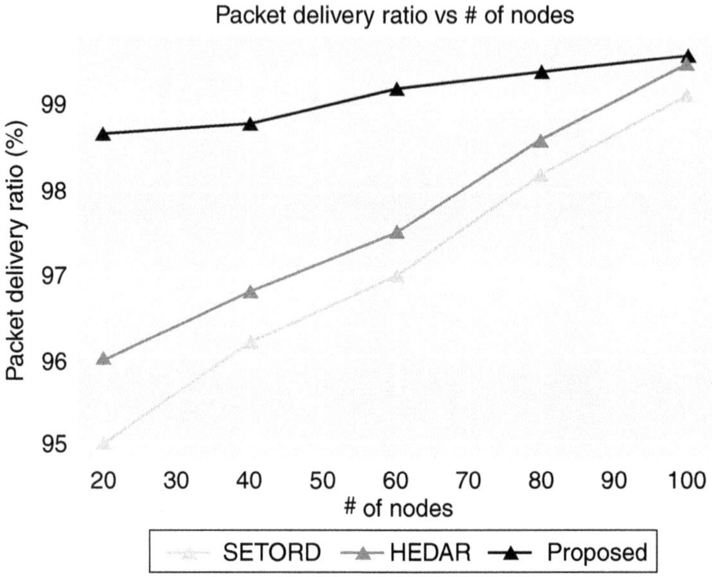

Figure 8.16 PDR comparison.

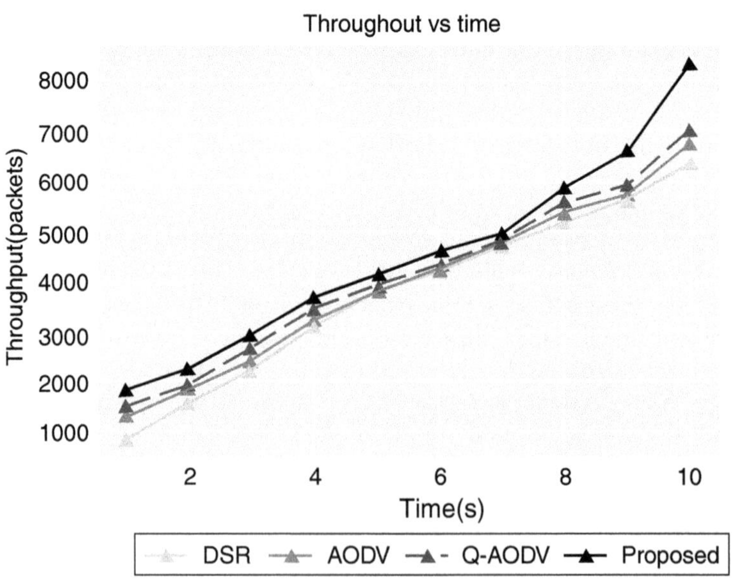

Figure 8.17 Throughput comparison.

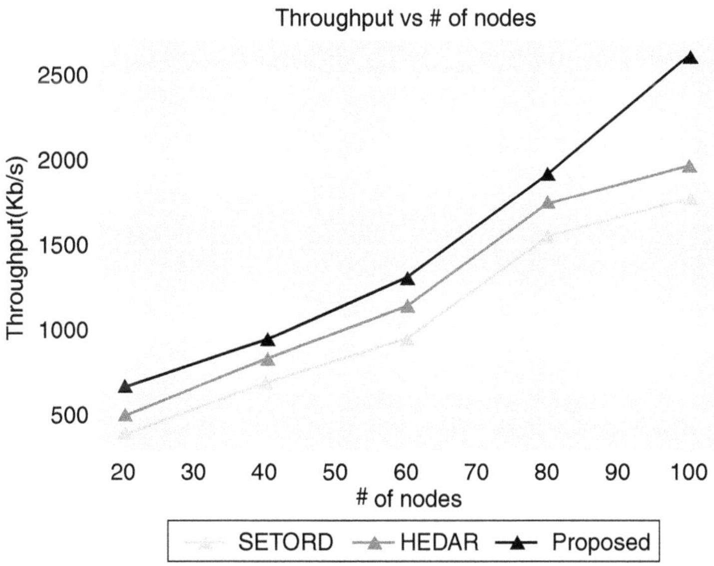

Figure 8.18 Throughput comparison.

demonstrates a significantly higher throughput when compared to previous technologies, ensuring an elevated delivery rate.

The proposed technique was evaluated in comparison to previous technologies, such as HEDAR and SETORD. The comparison diagram in Figure 8.18 illustrates the correlation between the number of nodes and throughput. The figure clearly illustrates a significant increase in throughput in comparison to the previous technique.

Figure 8.19 presents the results of comparing previous methodologies with the proposed techniques for assessing the number of packets dropped in the MANET. The graph demonstrates that the proposed method effectively attains the minimum packet loss rate.

Figure 8.20 illustrates the comparative analysis of the proposed network against prior networks regarding end-to-end latency for MANET data packets. The graph indicates that the proposed method enhances performance concerning the end-to-end delay of the packets.

Figure 8.21 presents a comparison of end-to-end latency and the number of nodes between the proposed technique and previous methods, including SETORD and HEDAR. The end-to-end delay demonstrates reduced effectiveness when compared to the prior technique, as illustrated by the figure. Figure 8.22 presents the resulting plot comparing the old and new routing overhead methods within the MANET framework. The graph demonstrates

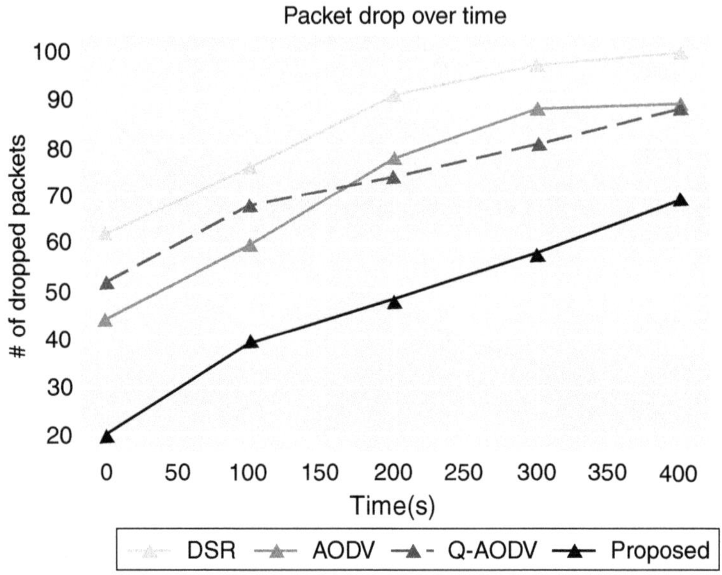

Figure 8.19 Comparison for packet dropped.

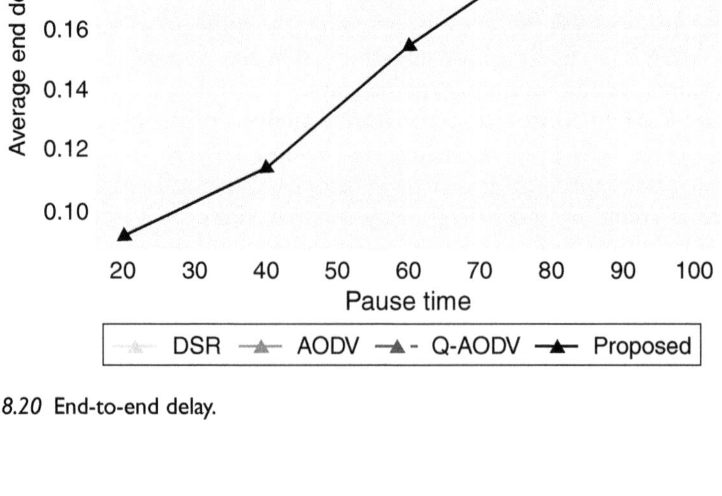

Figure 8.20 End-to-end delay.

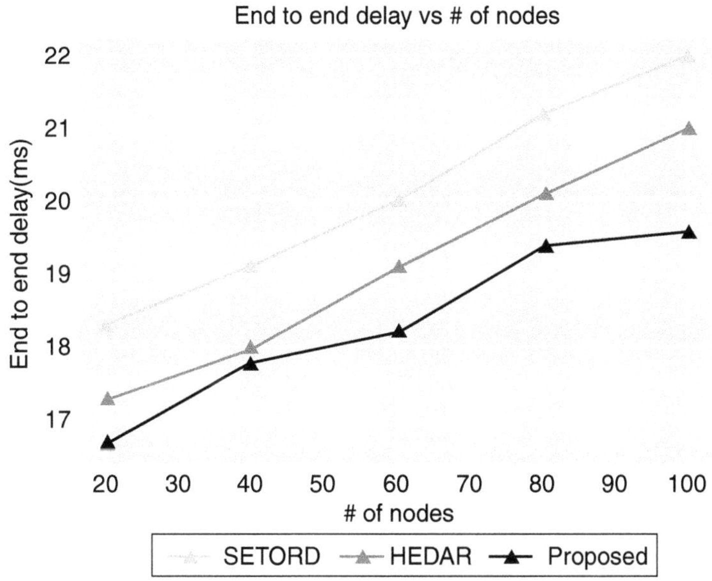

Figure 8.21 End-to-end delay.

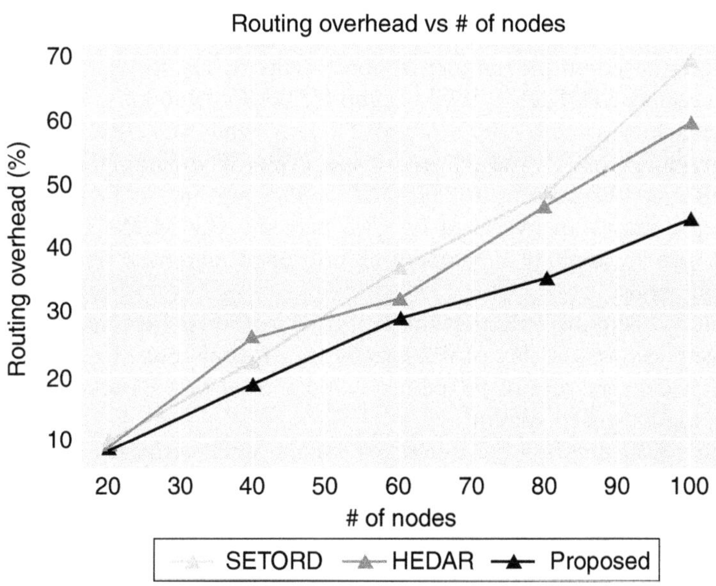

Figure 8.22 Comparison for routing overhead.

Table 8.4 Comparative analysis of the parameter

Methodologies	*Number of Nodes*	*PDR (%)*	*Throughput (kbps)*	*End-to-End Delay (ms)*	*Routing Overhead (%)*
HEDAR	20	517	96	17.3	9
	80	1763	98.6	20.1	45.2
	40	837	96.8	18	28.4
	100	1986	99.5	21	59.9
	60	1156	97.5	19.1	32.5
SETORD	20	395	95	18.3	10
	80	1564	98.2	21.2	47.9
	40	696	96.2	19.1	21.2
	100	1787	99.1	22	69.8
	60	956	97	20	38.3
Proposed	20	650	98.6	15	8.9
	80	1980	99.3	19.1	36.5
	40	960	98.8	17.8	19.6
	100	2650	99.6	19.5	44.2
	60	1350	99.1	18.2	30.1

that the proposed method effectively decreases the routing overhead within the Blockchain MANET.

Table 8.4 presents a comparative analysis of the parameters related to the prior approach and the proposed method. In comparing older methods such as SETORD, which attains a PDR of 99.1 percent, and HEDAR, with a ratio of 99.5 percent, it is essential to recognise that the characteristics also differ based on the number of nodes utilised. The throughput of 2,650 kb/s exceeds SETORD's 1,787 kb/s and HEDAR's 1986 kb/s. The proposed method achieves end-to-end delays of 19.5 ms, while SETORD and HEDAR exhibit delays of 22 and 21 ms, respectively. The routing overhead for SETORD is 69.8 percent, for HEDAR is 59.9 percent, while the proposed technique shows an overhead of 44.2 percent. The MANET method for Blockchain technology demonstrates improved efficiency relative to previous methods.

Table 8.5 demonstrates that the proposed method is expected to attain an average end-to-end delay of 0.20 seconds, a throughput of 8,500 packets, a packet delivery rate of 98 percent, and a total of 69 packets discarded within the proposed system.

Figure 8.23 presents the outcomes for the QoS settings. Enhanced efficiency was achieved through advancements in both throughput and PDR. The suggested method exhibits improved performance under conditions of minimal packet loss and reduced end-to-end latency. The proposed methodology yielded effective QoS results.

Figure 8.24 illustrates various types of attacks, including Blackhole, Grey, Warm, Sybil, Altered/Replayed or Spoofed Routing, Sinkhole, Hello Flood,

Table 8.5 Analysis of an existing with the proposed method

Methodologies	*PDR*	*Throughput*	*Number of packets dropped*	*Average end to end delay(s)*
DSR	91	6500	100	0.219
QAODV	96	7000	87	0.208
AODV	94	6800	89	0.21
Proposed	98	8500	69	0.2

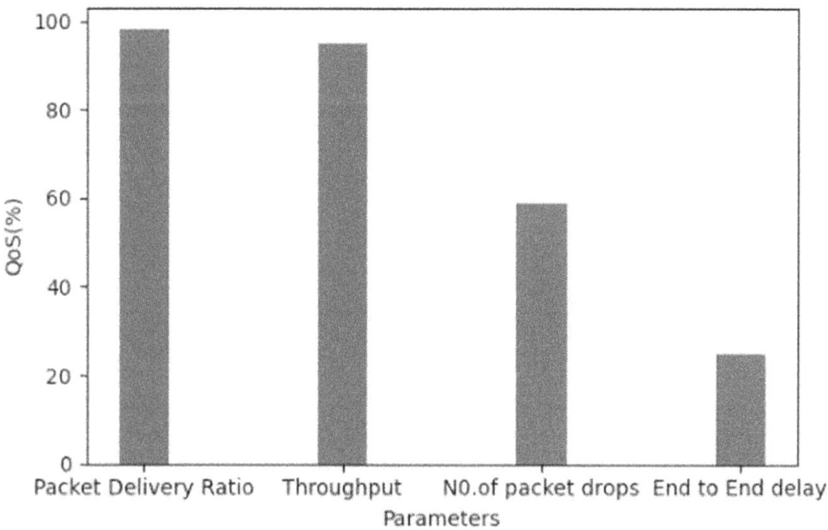

Figure 8.23 Analysis of various QoS parameters.

DOS-Apache2 Attack, R2L, and Probe Attacks. The graph illustrates that the proposed method achieves detection accuracies of 95, 97, 92, 94, 98, 96, and 98 percent for the following attacks: Black hole, Sybil, Altered / Replayed or Spoofed, Grey hole, Warm hole, Sinkhole, R2l, Probe, Hello flood, DOS attack-Apache2, in comparison to the prior method. Figure 8.25 presents a comparative analysis of the previous methods and the proposed method regarding attack detection percentages.

With percentages represented, Table 8.6 compares the proposed attack detection approach to the previous methods. The graph demonstrates that the proposed method achieves an attack detection rate of 95 percent, in contrast to AODV, which reaches 60 percent, and SETORD, which attains 75 percent. Table 8.5 provides a comparative analysis of the proposed attack detection method relative to existing methods, utilising percentage values for clarity. The graph indicates that the proposed method attains

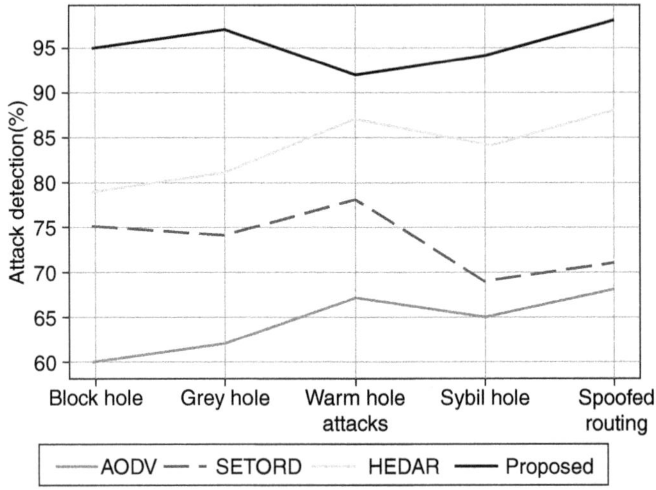

Figure 8.24 Attack detection.

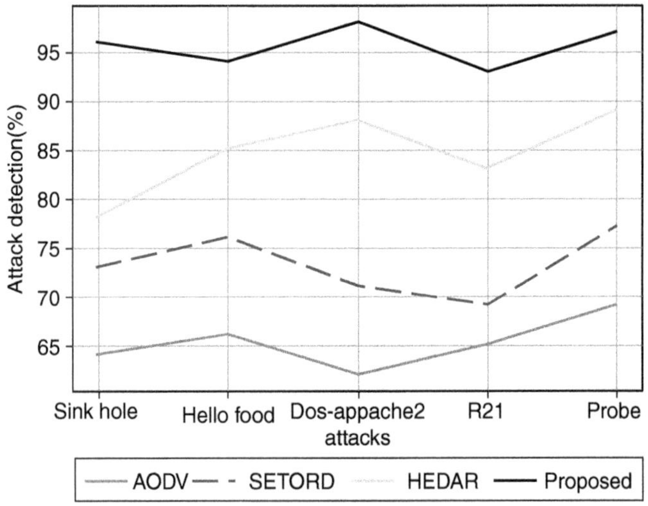

Figure 8.25 Attack detection.

a 95 percent attack detection rate for black hole attacks, whereas AODV records a 60 percent detection rate, SETORD achieves 75 percent, and HEDAR reaches 79 percent. The proposed technique achieves a detection rate of 97 percent, surpassing the performance of AODV, HEDAR, and SETORD, which demonstrate detection rates of 63, 74, and 82 percent

Table 8.6 Comparison of the prior methods and proposed method to Attack detection (%)

Methods / *Attacks*	*AODV*	*HEDAR*	*SETORD*	*Proposed*
Black hole	60	79	75	95
Sybil	65	84	69	94
Altered / Replayed or Spoofed	69	88	72	98
Grey hole	63	82	74	97
Warm hole	68	87	79	92
Sinkhole	64	78	74	96
R2I	65	83	69	93
Probe	69	89	73	97
Hello flood	66	85	76	94
DOS attack-Apache2	62	89	71	98

for Grey hole, respectively. The proposed method demonstrates an attack detection rate of 95 percent for black hole attacks, surpassing AODV's detection rate of 60 percent, HEDAR's rate of 79 percent, and SETORD's rate of 75 percent. The proposed method exhibits a detection rate of 92 percent for wormhole attacks, surpassing the detection rates of AODV, HEDAR, and SETORD, and HEDAR, which are 68, 87, and 79 percent, respectively. The proposed method demonstrates an attack detection rate of 94 percent, exceeding the detection rates of AODV at 65 percent, SETORD at 69 percent, and HEDAR at 84 percent when addressing the Sybil attack.

The proposed method achieves a detection rate of 98 percent for altered, replayed, or spoof attacks, which exceeds the detection rates of AODV at 69 percent, SETORD at 72 percent, and HEDAR at 88 percent. The proposed method outperforms AODV, SETORD, and HEDAR in detecting sinkhole assaults, with a rate of 96 percent compared to 64, 74, and 78 percent, respectively. The proposed technique demonstrates a detection rate of 94 percent for the Hello Flood Attack, exceeding the detection rates of AODV at 66 percent, SETORD at 76 percent, and HEDAR at 85 percent. The proposed technique demonstrates an attack detection accuracy of 98 percent against Apache2, surpassing AODV's detection rate of 62 percent, SETORD's rate of 71 percent, and HEDAR's rate of 89 percent. The proposed technique demonstrates a 93 percent attack detection rate in R21Attack, exceeding the detection rates of AODV at 65 percent, SETORD at 69 percent, and HEDAR at 83 percent. The proposed method achieves a 97 percent detection rate for Probe Attacks, exceeding the detection rates of AODV at 69 percent, SETORD at 73 percent, and HEDAR at 89 percent. The proposed system is expected to exhibit enhanced effectiveness when evaluated against existing systems and those without Blockchain integration.

8.5 CONCLUSION

In order to improve security within MANET, SRA with Blockchain is utilised. A lack of network integrity and an inadequate reliance management system are the root causes of the delay factor. Therefore, in order to resolve this problem, SRA advises employing a blockchain-based integrity system in order to bolster the existing level of security integrity. In addition to providing evidence that the reliance management system is effective, implement the promulgated reliance esteemed quadruplets condition in the Blockchain technology to improve the system's stability and scalability (Dba, 2021). In the technology known as the Blockchain, additional networks are connected in order to infect the QoS parameter caused by delay. The proposed SAR incorporating Blockchain technology will reduce the time delay factor. The results demonstrate that the suggested framework significantly enhances various parameters, including a high PDR, increased throughput, a lower number of dropped packets, and minimised end-to-end delay. This advancement contributes to the overall effectiveness of security in MANET by leveraging Blockchain technology and optimising data transmission based on performance metrics.

8.6 FURTHER RESEARCH

Blockchain technology and its potential applications in MANET will be the focus of our future projects. We intend to dig further into the ways Blockchain technology can protect MANETs from attacks and security threats. Additional research on hash consensus methods, ways to identify rogue nodes, best routes for data transmission, and ways to enhance performance between nodes could be undertaken to advance this design. More research is required in the fields of identity management, content distribution, and transmission, among others, since there are many questions and concerns that have not yet been resolved. In addition, the section primarily talks about how to use Blockchain technology to identify bad nodes and prevent the network from connecting to each other. To build a real public network, algorithms like proof-of-work (Nakamoto, 2008) delegated proof-of-stake, and routing paths that allow two nodes to reach a destination node are soon to be implemented.

REFERENCES

Abdelshafy, Mohamed A., and Peter J. B. King. 2016. "Resisting Blackhole Attacks on MANETs." *IEEE Annual Consumer Communications & Networking Conference*, January. https://doi.org/10.1109/ccnc.2016.7444935.

Anwar, Hasib. 2024. *"Blockchain Technology Explained: A Decentralized Ecosystem."* 101 Blockchains. February 19, 2024. https://101blockchains.com/blockchain-technology-explained/

Aswal, Pradeep. 2024. "Blockchain Nodes – Blockchain Council." *Blockchain Council*. June 15, 2024. www.Blockchain-council.org/blockchain/blockchain-nodes/.

Awan, Saba, Nadeem Javaid, Sameeh Ullah, Asad Ullah Khan, Ali Mustafa Qamar, and Jin-Ghoo Choi. 2022. "Blockchain Based Secure Routing and Trust Management in Wireless Sensor Networks." *Sensors* 22 (2): 411. https://doi.org/10.3390/s22020411

Bhagyalakshmi, None, and Amit Kumar Dogra. 2018. "Q-AODV: A Flood Control Ad-Hoc on Demand Distance Vector Routing Protocol." *2018 First International Conference on Secure Cyber Computing and Communication (ICSCCC)* 8 (December): 294–99. https://doi.org/10.1109/icsccc.2018.8703220

Borkar, Gautam M., and A. R. Mahajan. 2016. "A Secure and Trust Based On-demand Multipath Routing Scheme for Self-organized Mobile Ad-hoc Networks." *Wireless Networks* 23 (8): 2455–72. https://doi.org/10.1007/s11276-016-1287-y

Borkar, Gautam M., and Anjali R. Mahajan. 2018. "Security Aware Dual Authentication Based Routing Scheme Using Fuzzy Logic with Secure Data Dissemination for Mobile Ad-hoc Networks." *Journal of Applied Security Research* 13 (2): 223–49. https://doi.org/10.1080/19361610.2017.1387737

Careem, Maqsood Ahamed Abdul, and Aveek Dutta. 2020. "Reputation Based Routing in MANET Using Blockchain." *2020 International Conference on COMmunication Systems & NETworkS (COMSNETS)*, January, 1–6. https://doi.org/10.1109/comsnets48256.2020.9027450

Cordova, David, Alexandre Laube, Thi-Mai-Trang Nguyen, and Guy Pujolle. 2020. "Blockgraph: A Blockchain for Mobile Ad Hoc Networks." *2020 4th Cyber Security in Networking Conference (CSNet)*, October, 1–8. https://doi.org/10.1109/csnet50428.2020.9265532

Dba, Geraldo Vasquez. 2021. "An Introduction to Blockchain." *The CPA Journal*. August 18, 2021. www.cpajournal.com/2021/08/18/an-introduction-to-blockchain/

Dilli, Ravilla, and Putta Chandra Sekhar Reddy. 2016. "Implementation of Security Features in MANETs Using SHA-3 Standard Algorithm." *2016 International Conference on Computation System and Information Technology for Sustainable Solutions (CSITSS)* 1 (October): 455–58. https://doi.org/10.1109/csitss.2016.7779410

Dongarwar Kishor, and Ghodichor Nitesh. 2017. "Review on Black hole Attack Detection in Wireless MANET." *International Journal of Engineering Research in Computer Science and Engineering (IJERCSE)*, 4 (3): 1–5.

El-Bashary, M., A. Abdelhafez, W. Anis, and Dept. of Comm. and Electronics Ain Shams University, Dept. of Comm. and Electronics Ain Shams University, and Dept. of Comm. and Electronics Ain Shams University. 2015. "A Comparative Study of Group Key Management in MANET." *International Journal of Engineering Research and Applications* 5 (8), (Part 4): 85–94. www.ijera.com/papers/Vol5_issue8/Part%20-%204/K58048594.pdf

Ethereum. 2021. "Ethereum Whitepaper." Ethereum.Org. https://ethereum.org/en/whitepaper/.

GeeksforGeeks. 2021. "Blockchain Block Hashing." *GeeksforGeeks*. November 26, 2021. www.geeksforgeeks.org/blockchain-block-hashing/

Geroni, Diego. 2023. "Blockchain Nodes: An In-Depth Guide." 101 *Blockchains*. May 9, 2023. https://101blockchains.com/blockchain-nodes/

Ghodichor, Nitesh, Raj Thaneeghaivl V., Varsha Namdeoe, and Gautam Borkar. 2022. "Secure Routing Protocol against Internal and External Attack in MANET." *The International Conference on Emerging Trends in Artificial Intelligence and Smart Systems*, THEETAS 2022, January. https://doi.org/10.4108/eai.16-4-2022.2318163

Iredale, Gwyneth. 2023. "What Are the Different Types of Blockchain Technology?" *101 Blockchains*. May 19, 2023. https://101blockchains.com/types-of-blockchain/

V. S., Janani, and M. S. K., Manikandan. 2018. "Efficient Trust Management with Bayesian-Evidence Theorem to Secure Public Key Infrastructure-based Mobile Ad Hoc Networks." *Journal on Wireless Communications and Networking* 25. https://doi.org/10.1186/s13638-017-1001-5

Jothi, J. Sathiya. 2018. "A Survey on Trust Management for Mobile Ad Hoc Networks." *IJERT*, April. https://doi.org/10.17577/IJERTCONV4IS24012

Jothi, J. Sathiya, S. Senthilkumar, and B. Rajesh. 2016. "A Survey on Trust Management for Mobile Ad Hoc Networks, International Journal of Engineering Research & Technology (IJERT) AMASE." 4 (24).

Khullar, Kashish. 2020. "Decentralized and Secure Communication Architecture for FANETs Using Blockchain." *Procedia Computer Science* 173 (January): 158–70. https://doi.org/10.1016/j.procs.2020.06.020

Madhusudhanan, B., S. Chitra, and C. Rajan. 2015. "Mobility Based Key Management Technique for Multicast Security in Mobile Ad Hoc Networks." *The Scientific World Journal* 2015 (January): 1–10. https://doi.org/10.1155/2015/801632

Mouchfiq, N., A. Habbani, and C. Benjbara. 2019. "Blockchain Security in MANETs". *World Academy of Science, Engineering and Technology, Open Science Index 154, International Journal of Computer and Information Engineering* 13 (10), 542–46.

Mukhedkar, Moresh Madhukar, and Uttam Kolekar. 2020. "E-TDGO: An Encrypted Trust-based Dolphin Glowworm Optimization for Secure Routing in Mobile Ad Hoc Network." *Communication System* 33, e4252, 1074–5351. https://doi.org/10.1002/dac.4252

Nakamoto, S. 2008" Bitcoin". *A Peer-to-Peer Electronic Cash System.bitcoin.org*.

Patil, Vaishali, P. Fulare, S. S. Patil, and N. Ghodichor. 2014. "A Review on an Unobservable Secure Routing Protocol with Wormhole Attack Prevention in Manet." *International Journal of Current Engineering and Technology* 4 (3), 1930–32. http://inpressco.com/category/ijcet

Perumal, V. Saravana, P. Sathya, Kumar E. Rajesh, and Parvez A. Mohamed. 2020. "Design of an Internet of Things Approach for Industrial Pollution Monitoring." *International Journal of Engineering Research & Technology (IJERT) ECLECTIC* 8 (7). https://doi.org/10.17577/IJERTCONV8IS07013

Pilankar, Pooja, and Puja Padiya. 2016. "Trust Based Security in Manet." *Journal-article. International Journal of Research in Engineering and Technology* 5 (02), 11–17. https://ijret.org/volumes/2016v05/i02/IJRET20160502002.pdf

Prasath, N., and J. Sreemathy. 2018. "Optimized Dynamic Source Routing Protocol for MANETs." *Cluster Computing* 22 (S5): 12397–409. https://doi.org/10.1007/s10586-017-1638-1

Ramezan, Gholamreza, and Cyril Leung. 2018. "A Blockchain-Based Contractual Routing Protocol for the Internet of Things Using Smart Contracts." *Wireless Communications and Mobile Computing* 2018 (January): 1–14. https://doi.org/10.1155/2018/4029591

Sandeep, None. 2012. "Comparing the Impact of Black Hole and Gray Hole Attacks in Mobile Adhoc Networks." *Journal of Computer Science* 8 (11): 1788–802. https://doi.org/10.3844/jcssp.2012.1788.1802

Sen, Jaydip, and Innovation Lab, Tata Consultancy Services Ltd. 2010. "A Distributed Trust and Reputation Framework for Mobile Ad Hoc Networks." *Journal-Article Innovation Lab*, Tata Consultancy Services Ltd. https://arxiv.org/abs/1012.2519

Singh, Sachchidanand, and Nirmala Singh. 2016. "Blockchain: Future of Financial and Cyber Security." *International Conference on Contemporary Computing and Informatics*, December. https://doi.org/10.1109/ic3i.2016.7918009.

Sinwar, Deepak, Nisha Sharma, Sunil Kumar Maakar, and Sudesh Kumar. 2020. "Analysis and Comparison of Ant Colony Optimization Algorithm with DSDV, AODV, and AOMDV Based on Shortest Path in MANET." *Journal of Information and Optimization Sciences* 41 (2): 621–32. https://doi.org/10.1080/02522667.2020.1733193

Srilakshmi, Uppalapati, Saleh Ahmed Alghamdi, Veera Ankalu Vuyyuru, Neenavath Veeraiah, and Youseef Alotaibi. 2022. "A Secure Optimization Routing Algorithm for Mobile Ad Hoc Networks." *IEEE Access* 10 (January): 14260–69. https://doi.org/10.1109/access.2022.3144679

Thiagarajan, R., R. Ganesan, V. Anbarasu, M. Baskar, K. Arthi, and J. Ramkumar. 2021. "Retracted Article: Optimised with Secure Approach in Detecting and Isolation of Malicious Nodes in MANET." *Wireless Personal Communications* 119 (1): 21–35. https://doi.org/10.1007/s11277-021-08092-0.

Usha, M. S., and K. C. Ravishankar. 2021. "Implementation of Trust-Based Novel Approach for Security Enhancements in MANETs." *SN Computer Science* 2: 257. https://doi.org/10.1007/s42979-021-00628-2

Veeraiah, Neenavath, Osamah Ibrahim Khalaf, C. V. P. R. Prasad, Youseef Alotaibi, Abdulmajeed Alsufyani, Saleh Ahmed Alghamdi, and Nawal Alsufyani. 2021. "Trust Aware Secure Energy Efficient Hybrid Protocol for MANET." *IEEE Access* 9 (January): 120996–5. https://doi.org/10.1109/access.2021.3108807

Wessling, Florian, and Volker Gruhn. 2018. "Engineering Software Architectures of Blockchain-Oriented Applications." *IEEE International Conference on Software Architecture Companion*, April, 45–46. https://doi.org/10.1109/icsa-c.2018.00019

Zach, Rynes and ChainLinkGod. 2022. "What Everyone Gets Wrong About Blockchains." *Chainlink Blog*. October 7, 2022. https://blog.chain.link/blockchain-trust-model

Chapter 9

Blockchain in healthcare

Case studies and applications

Kaikashan I. Siddavatam, Ojas Arun Yadav, and Vedant Ganesh Padwal

9.1 INTRODUCTION

The healthcare sector is a rapidly growing sector in the world. The Blockchain is gradually becoming a disruptive innovation that can address the major inefficiencies in the world [1]. Hospitals and health care organizations everywhere are facing problems such as data breaches, fraudulent activities, lack of patient control, and the double inefficiency of medical records and healthcare supply chain management. The open ledger has several characteristics that make it possible to supply a decentralized, tamper-proof, and traceable infrastructure for data management.

9.1.1 The challenges in healthcare

The most important challenge is healthcare data breaches, and in 2022 alone, there were more than 50 million exposed health records attributed to cybersecurity compromise. Centralized paradigms for data storage are extremely vulnerable to cyber-attacks, modification, and unauthorized access [6]. Moreover, counterfeit drugs are estimated to cost the global pharmaceutical industry roughly $200 billion per year [2][3], with considerable risks to patients. On top of this challenge, the rise of chronic illnesses like diabetes and hypertension are contributing to increasing demands for remote medical monitoring in real time. Most current systems are inefficient, highly prone to error, and lack security.

9.1.2 Blockchain's role in healthcare transformation

Blockchain technology helps to solve these problems through decentralizing storage of data, secure and transparent transactions, and greater patient control over health data [3]. It has critical functional features such as immutability, consensus mechanisms, and cryptographic security, which make it fundamentally viable for healthcare applications [4][5]. This chapter will introduce the use of Blockchain in healthcare and will review its outcomes

DOI: 10.1201/9781003584728-9

through real-world examples and experimentation to evaluate whether it can be effective at solving or ameliorating some of the challenges while increasing the effectiveness and trust of the transactions among stakeholders in healthcare.

9.2 LITERATURE REVIEW

9.2.1 Blockchain for secure patient data management

It is necessary that we protect patient data and maintain their privacy in the healthcare system. At the moment, electronic health records (EHRs) are kept in centralized databases, which puts them at the risk of being compromised. The use of Blockchain technology has been hailed as the only solution that enables decentralized storage of patient information, thereby ensuring the security of the patients and the safety of their data.

MedRec: The first example MedRec is an EHR management system that works on a block chain, created by researchers at the Massachusetts Institute of Technology (MIT) [2]. The system enables patients to manage their health data, including the ability to authorize specific providers or institutions to access their data. The MedRec system uses the smart contracts of Ethereum for automating access permissions and for recording provides interactions with patients, which results in providing transparency and accountability.

Sophisticated cryptographic mechanisms block chain technologies pertain to advanced cryptographic mechanisms, such as the following:

- *Zero-knowledge proofs* (ZKPs) [5][6]: Allow for verification of data without revealing its content to an external party, thus maintaining privacy and allowing transparency[7].
- *Homomorphic encryption*: The technology makes it possible to calculate on encrypted data and thus allows the data providers to carry out the analytics on the patient data without exposing the sensitive parts of the data[8].
- Impact bBlockchain creates trust and accountability for patient data management in a decentralized nature. A direct result of Blockchain implementation is the reduction or elimination of intermediaries, which reduces operational costs, simplifies the authorization process, and grants patients control of their data[9][10].

9.2.2 Blockchain in pharmaceutical supply chain management

The pharmaceutical industry faces challenges including counterfeit drugs, inefficiencies in supply chain logistics, and a lack of transparency[11].

Blockchain can be thought of as a solution to provide end-to-end traceability, verify drug authenticity, and create efficiencies within the supply chain.

1. *MediLedger*: Addressing counterfeit drugs, MediLedger is a Blockchain platform designed to track pharmaceutical products throughout the supply chain to ensure land traceability. The history of all transactions from manufacturer to distributor is securely logged on a privy, immutable ledger. MediLegder also connects Internet of Things (IoT) devices, such as radiofrequency identification (RFID) tags and temperature sensors, to track environmental conditions and keep products in compliance with storage requirements. IoT in the supply chain IoT devices will improve the efficiency of any Blockchain systems with real-time data.
2. *Temperature sensors*: Track storage conditions for vaccines and biologics.
3. *RFID tags*: Provides unique digital identity that allows tracking of individual products.

Repeated pilot programs have demonstrated that the use of MediLedger has decreased counterfeit drugs instances by 60 percent, improved compliance, and builds stakeholder confidence.

9.2.3 Blockchain for remote patient monitoring and integrating IoT

The combination of Blockchain technology and IoT devices has a major impact on remote patient monitoring. Wearable devices like fitness trackers or glucose monitors can collect health data in real time[8][12]. This data is then shared and securely stored using Blockchains.

Smart contracts enable certain functions to happen in the healthcare field, including:

- The sending of alerts when a patient's vitals are in an abnormal range
- The sending of reminders for when a patient should take medications and how much they should take

The sending of immediate notifications of emergencies to healthcare providers or caregivers.

Scalability problems and solutions: As they provide a real-time basis for data collection, IoT networks can aggregate considerable amounts of data, leading to scaling problems for Blockchain systems[13][14].

Hybrid storage frameworks are one alternative that can provide both on-chain and off-chain storage of data:

- Important data is stored on a Blockchain.
- Other data is stored on a cloud data-based system.

9.2.4 Privacy-preserving Blockchain models in health care

Privacy is a major impediment to Blockchain use in healthcare. It is important that privacy and transparency coexist in healthcare because transparency will usually expose sensitive metadata whether intended or not. Privacy-preserving techniques allow for compliance with the applicable privacy regulations without sacrificing security[7][15].

1. **Techniques for privacy preservation**

- *ZKPs*: a technique for validating the authenticity of data without exposure to its contents.
- *Ring signatures*: a technique for authenticating someone without revealing their identity.
- *Private transactions*: a technique for sharing secure data only with selected parties.

2. Compliance with regulation
The above techniques are compliant with international health care regulations such as the General Data Protection Regulation (GDPR) and Health Insurance Portability and Accountability Act (HIPAA) as prescribed for the secure sharing and privacy protections of patient data.

9.3 METHODOLOGY

9.3.1 Integration of IoT and Blockchain

The integration of IoT and BlockchainBlockchain in healthcare is complex, but it is one that ensures secure, real-time collection, transmission, and storage. This section will elaborate further on the detailed steps and techniques used in this integration.

Proposed method diagram: Figure 9.1 represents the process of integration of the Blockchain and IoT healthcare collection of health data using wearable sensors IoT devices, such as continuous glucose monitors (CGMs) or fitness trackers, which are essential for seamlessly collecting real-time health metrics. These devices are continuously monitoring different indicators of health, such as:

- Heart rate: Monitoring heart beats per minute and cardiovascular health.

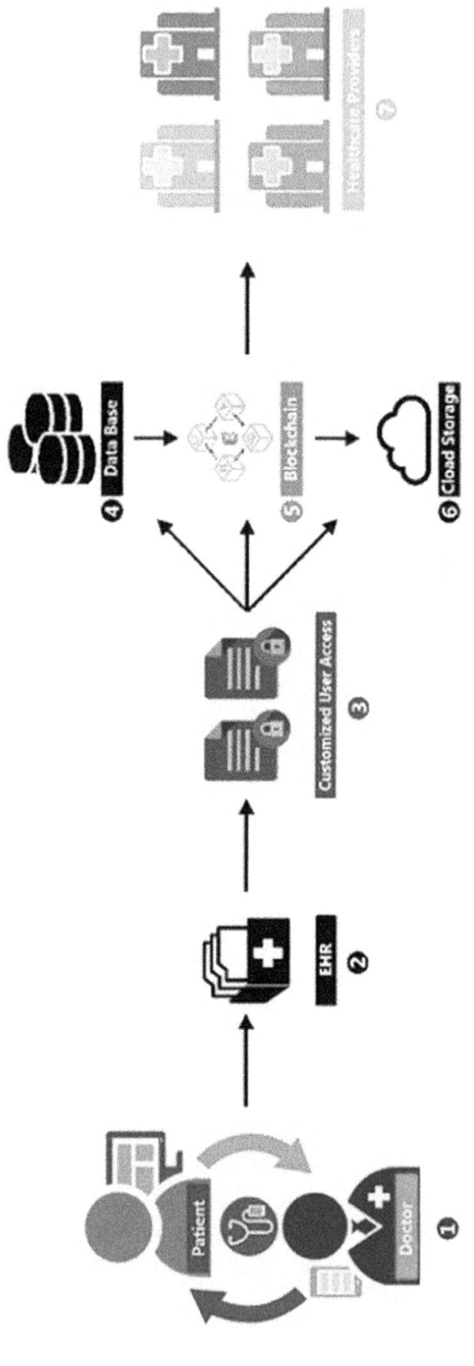

Figure 9.1 Blockchain technology in healthcare.

- Blood sugar levels: Management of diabetes through CGMs.
- Physical activity: Counting steps taken, distance, and calories burned.

These metrics hold valuable data in understanding a patient's health condition as well as health trends over time.

Secure transmission of data: Once collected, critical data collected through IoT devices is securely transmitted to mitigate threats of unauthorized access and uphold patient privacy. Below are some key techniques for using secure data transmission:

- *AES-256 encryption*: Advanced Encryption Standard (AES) with a 256-bit key is used to encrypt data. AES ranks as one of the most secure standards for data encryption, which is particularly effective in protecting data that are being transferred.
- *Device authentication*: IoT devices and the network nodes are able to authenticate with each other using a digital certificate, for example, an X.509 certificate, that would ensure that only authorized devices are allowed to communicate with the trusted node within the network.
- *Secure communication protocols*: Hypertext Transfer Protocol Secure (HTTPS), Message Queuing Telemetry Transport (MQTT) with Transport Layer Security, and Constrained Application Protocol (CoAP) are implemented to establish secure channels of communication between IoT devices and back-end servers.

Automation through Smart Contracts: Smart contracts are self-executing codes that run on a Blockchain and enforce rules and processes automatically. Smart contracts can automate several tasks in healthcare, including:

- *Alerts*: Smart contracts can automatically trigger alerts when a patient's vitals reach an abnormal state.
- *Medication reminders*: Smart contracts can send reminder notifications automatically to patients to take their medications, which will improve adherence.
- *Scheduling follow-up appointments*: Smart contracts can schedule follow-up appointments automatically from health data, notifying both the patient and health care provider of the time.

This type of automation will increase efficiency, decrease human error, and allow for timely interventions.

Hybrid storage models: The amount of data generated from IoT devices can be overwhelming, so hybrid storage models are used to balance the

Blockchain immutability benefit with scalable storage. Hybrid storage includes:

- *On-chain*: Critical health data like medical records and patient identifiers are stored on the Blockchain. This allows for immutability and transparency.
- *Off-chain*: Less critical data like raw IoT sensor readings can be stored in off-chain systems like cloud-based storage. This hybrid storage will allow optimization of performance at scale.

9.3.2 Data processing and analysis

After data is reliably obtained and stored, it needs to be analyzed so we can begin deriving knowledge from it. This is a multi-step process needs to occur:

- *Data cleaning*: This implies correcting and/or removing inaccurate, incomplete, or irrelevant data entries to have a quality data set.
- *Data transformation*: This involves editing the raw data to make it ready for analysis. It may consist of normalizing values, aggregating data points, or just using some stats.
- *Data integration*: This is the combining multiple sources of data together into an analysis-ready dataset (e.g. EHRs, IoT data, third-party databases).
- *Analytical models*: The core of learning information from our analyses is using machine learning algorithms and statistical models to find patterns, trends, and correlations in data.

All of these processes will allow healthcare providers to make decisions, provide better patient outcomes, streamlined care, and better operational efficiency.

9.3.3 Compliance and regulatory considerations

The deployment of Blockchain and IoT technology in healthcare, which is successful, requires a thorough understanding of various regulations and standards that manage the confidentiality and security of data related to patient care. These are the core principles that form the basis for the rest:

- *HIPAA compliance*: It is vital that any data handling that includes patient health information should observe the requirements given by HIPAA, which regulates health insurance accountability and portability.

- *GDPR compliance*: While that statement is made about HIPAA, healthcare providers handling or providing care to patients in the European Union must take precaution to follow the newly enacted GDPR to protect patient data and privacy rights.
- *Interoperability standards*: Using standards developed to encourage interoperability in healthcare, such as HL7 (Health Level 7) or FHIR (Fast Healthcare Interoperability Resources), secure healthcare data should be able to flow freely between health delivery systems.

In integrating Blockchain and IoT in a manner that observes secure collection, transmission, and storage of data, processing of data provided along with HIPAA and GDPR regulations will expand the capabilities of the technology and the healthcare industry to improve patient care and health organizational efficiency.

This methodology cultivates a multi-faceted approach to implementing Blockchain and IoT in the healthcare industry to meet the technical and regulatory obstacles while maximizing improvements toward the benefit of the patient and care provider.

9.4 CASE STUDIES

9.4.1 HealthChain: Remote patient monitoring

Motivation and objective: HealthChain was developed to address the challenges of chronic disease management by integrating IoT devices with Blockchain technology. The primary goal is to provide secure, real-time monitoring of patients with diabetes, hypertension, and cardiovascular conditions. This system ensures continuous health tracking, automates responses through smart contracts, and improves patient outcomes while reducing healthcare costs.

BlockchainAs shown in Figure 9.1, a detailed illustration of the integration of Blockchain and IoT in HealthChain is presented, and the complete working of the method is explained below.

1. *Data collection*: Patients use IoT devices to monitor health metrics such as glucose levels and blood pressure.
2. *Data encryption*: The collected data is encrypted using AES-256 to ensure privacy and security.
3. *Data transfer*: Encrypted data is sent to the Blockchain network via secure communication protocols.
4. *Smart contracts*: Trigger alerts and automate responses when abnormal values are detected. Smart contracts also send medication reminders and schedule follow-up appointments.

5. *Data storage*: Critical health data is stored on the Blockchain (on-chain), while less sensitive data is stored in cloud storage systems (off-chain).
6. *Data access and insights*: Healthcare providers access the data through a secure interface. Machine learning models transform and mine the data to provide actionable insights, helping doctors make informed decisions and improve patient outcomes.

Role of Blockchain and IoT

- *IoT*: IoT devices serve as the backbone for collecting real-time health data. These devices, such as wearable sensors (e.g., fitness trackers and glucose monitors), continuously monitor vital health metrics like blood pressure, glucose levels, heart rate, and physical activity. The constant stream of data allows for real-time analysis, enabling healthcare providers to identify abnormalities, track trends, and make proactive interventions. IoT devices ensure that patient health is consistently monitored without requiring physical visits, promoting remote healthcare solutions.
- *Blockchain*: Blockchain technology plays an important role in the security, reliability, and privacy of the collected health data. Key features include:
    - *Data security and privacy*: Blockchain uses encryption techniques like AES-256 to secure patient data. Once data is stored on the Blockchain, it cannot be altered, ensuring immutability and reliability.
    - *Decentralized storage*: By distributing data across a Blockchain network, the risk of single-point failures is eliminated. Data remains accessible and protected from unauthorized access.
    - *Smart contracts*: These self-executing scripts automate critical healthcare processes, such as triggering alerts when health parameters deviate from normal ranges, sending reminders for medication adherence, and scheduling follow-up appointments. Automation reduces human errors and ensures timely interventions.
    - *Transparency and trust*: Blockchain offers an open and verifiable log of data transactions of any kind, thereby offering a guarantee of trust to patients as well as healthcare providers and the latter's stakeholders. Patients are in charge of deciding who should have access to their data which is a more efficient way of managing consent.

Together, the integration of IoT and Blockchain enhances healthcare management by providing a secure, real-time, and automated solution for chronic disease monitoring and intervention.

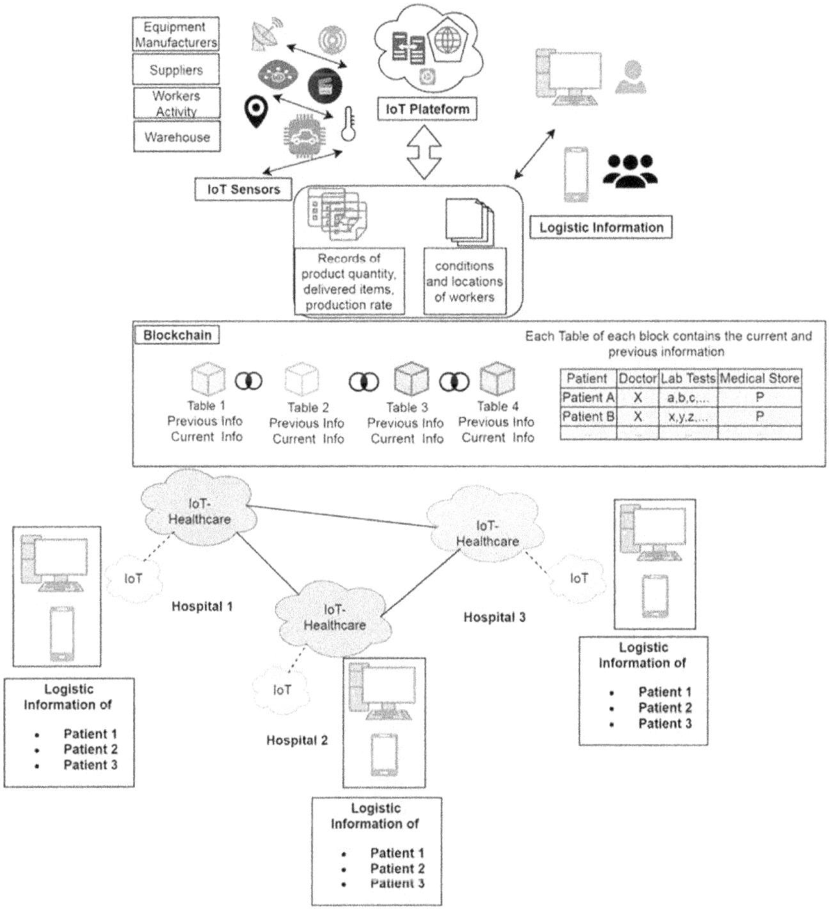

Figure 9.2 Integration of Blockchain and IoT.

Reading and graphs: To illustrate the effectiveness of HealthChain, we conducted a six-month pilot program involving 200 patients with chronic conditions. Figure 9.2 shows the improvements in patient outcomes and reduction in healthcare costs.

HealthChain has demonstrated the ability to boost patient outcomes by 30 percent and lower healthcare expenses through trial initiatives (Figure 9.3). The system's capacity to deliver up-to-the-minute data and carry out reaction procedures without human intervention greatly increased the effectiveness of long-term disease management. Besides cutting down on the number of medical emergencies, patients' well-being has been elevated since health issues are addressed before they escalate.

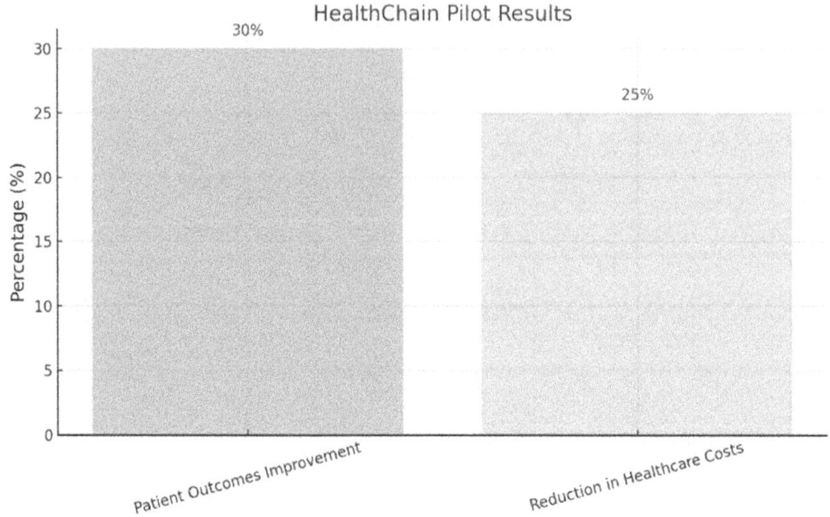

Figure 9.3 Health chain pilot results.

9.5 RESULTS AND EXPERIMENTS

9.5.1 Experiment design

The effectiveness of Blockchain in healthcare was evaluated through a comprehensive clinical trial involving 1,000 participants across three major hospitals. This experiment focused on assessing Blockchain's ability to securely handle sensitive health data, streamline remote monitoring, and improve response times in critical scenarios.

The primary objective was to integrate Blockchain and IoT to enable secure, real-time data collection, storage, and sharing for chronic disease management. This included evaluating system performance, user satisfaction, and scalability.

Experiment Setup

1. Devices used:
   - CGMs for diabetic patients.
   - Blood pressure monitors for hypertensive patients.
   - Wearable fitness trackers for general health monitoring.
2. Data collection and transmission:
   - IoT devices collected real-time health data and transmitted it using secure Bluetooth and Wi-Fi protocols.

3. Blockchain storage:
   - *On-chain storage*: Critical health events and alerts were stored immutably on the Blockchain.
   - *Off-chain storage*: Non-essential data, such as daily activity logs, was stored in cloud-based systems.
4. Evaluation metrics:
   - *Data integrity*: Ensured through cryptographic hashing and Blockchain immutability.
   - *Security*: Assessed by tracking unauthorized access attempts and tampering incidents.
   - *Response time*: Measured the time taken for alerts to be processed and acted upon.
   - *Usability*: Feedback was collected from both participants and healthcare providers.

Implementation: The Blockchain network was implemented using Hyperledger Fabric, a permissioned Blockchain platform. It included three key entities: hospitals, patients, and researchers. IoT devices were authenticated through X.509 certificates, and data encryption was applied using AES-256.

9.5.2 Results

The experiment produced promising results, showcasing Blockchain's and IoT's specific contributions to improving healthcare outcomes. The following highlights key tasks performed by the system, along with evaluation metrics supported by model results and graphs.

9.5.2.1 Specific tasks performed

1. *Blockchain's role*:
   - *Immutable storage*: Critical health alerts (e.g., glucose spikes or blood pressure abnormalities) were stored immutably on the Blockchain, ensuring no data manipulation.
   - *Secure sharing*: Health data was securely shared between patients, hospitals, and researchers, improving collaboration in chronic disease management.
2. *IoT's contribution*:
   - *Continuous monitoring*: IoT devices (CGMs, blood pressure monitors, and fitness trackers) provided real-time health readings, such as glucose levels, blood pressure values, and step counts.
   - *Automated alerts*: IoT devices triggered alerts when readings crossed predefined thresholds (e.g., glucose >180 mg/dL or blood pressure >140/90 mmHg).

- *Data transmission*: Secure Bluetooth and Wi-Fi protocols ensured uninterrupted real-time data flow to healthcare systems.

9.5.2.2 Evaluation metrics

Below are the detailed evaluation metrics along with visual representation through graphs:

1. Data integrity
   - *Outcome*: 100 percent of health data stored on the Blockchain remained tamper-proof.
   - *Metric*: Cryptographic hashing and immutability maintained zero incidents of data modification.
   - Graph (Figure 9.4)
2. Security
   - *Outcome*: Unauthorized access attempts were successfully blocked.
   - *Metric*:
     - Total unauthorized access attempts: 72
     - Success rate of encryption and authentication protocols: 100 percent
   - Graph (Figure 9.5)

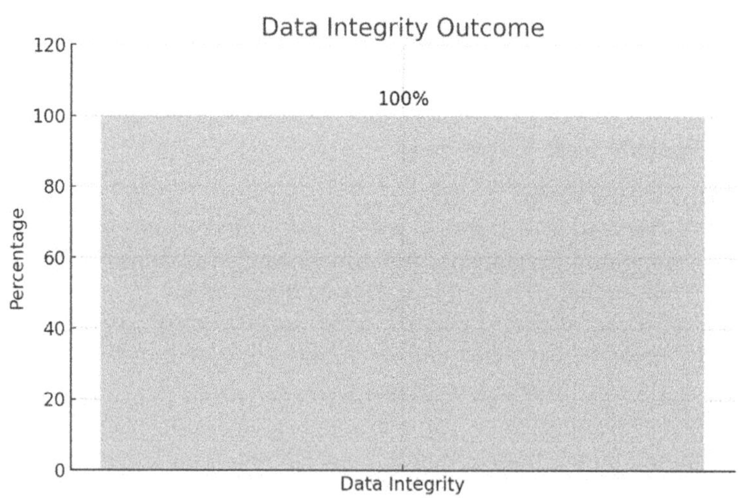

Figure 9.4 Data integrity outcome.

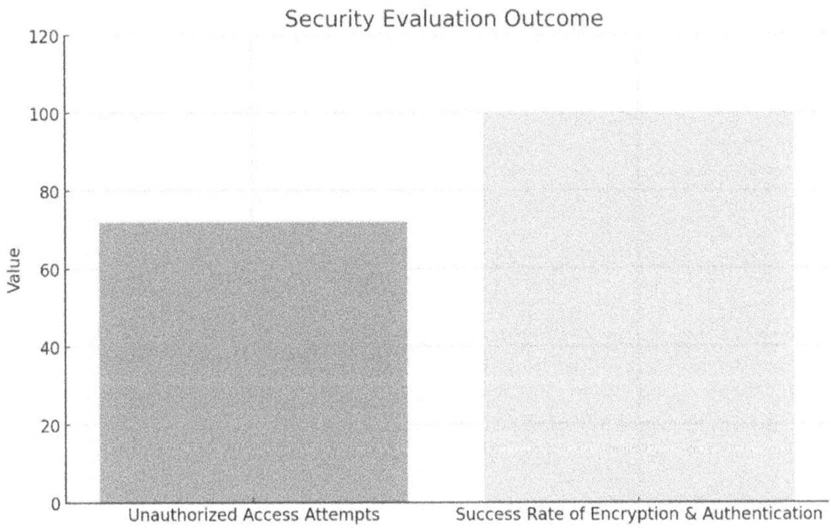

Figure 9.5 Security evaluation outcome.

3. Response time
   - *Outcome*: Smart contracts significantly reduced response time.
   - *Metric*:
     - Average alert delivery time: 10 seconds
     - Improvement: 40 percent reduction compared to traditional systems (previous response: 17 seconds).
   - Graph (Figure 9.6)
4. Scalability
   - *Outcome*: System handled 1 million+ data points without degradation.
   - *Metric*:
     - Data processed: 1,045,000 data points
     - Average latency per transaction: 3.2 milliseconds
   - Graph (Figure 9.7)
5. User satisfaction
   - *Outcome*: High satisfaction reported among participants and healthcare providers.
   - *Metric*:
     - Participant satisfaction: 95 percent
     - Provider satisfaction: 90 percent
   - Graph (Figure 9.8)

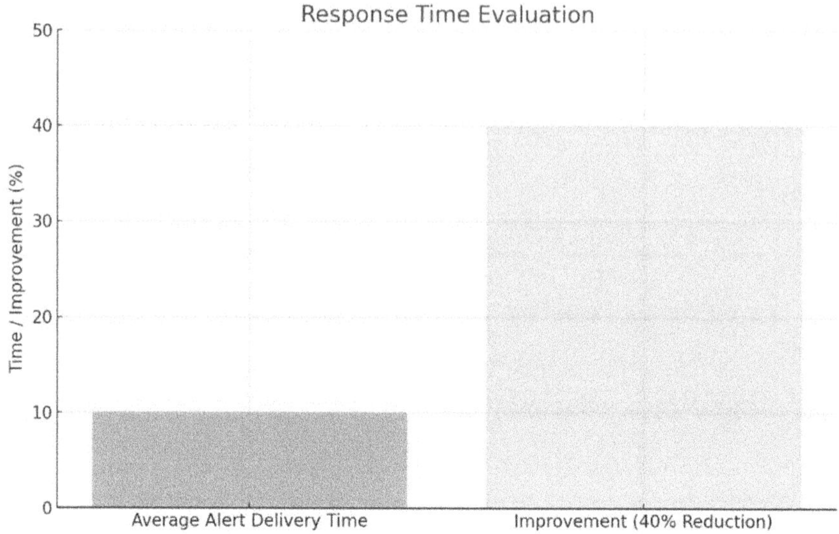

Figure 9.6 Response time evaluation.

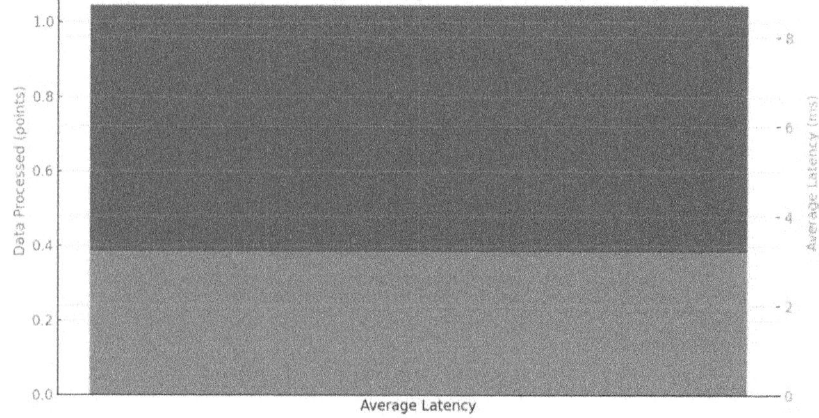

Figure 9.7 Scalability metrices.

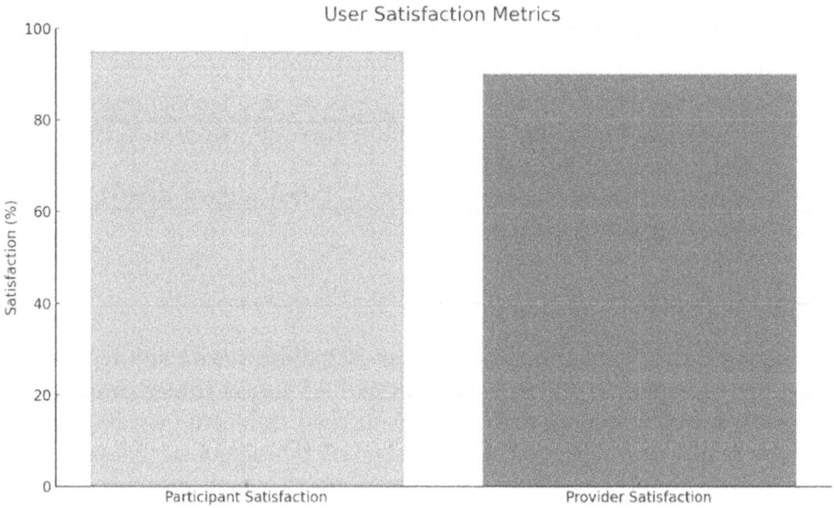

Figure 9.8 User satisfaction metrics.

9.5.2.3 *Impact on healthcare outcomes*

- Hospitalizations reduced by 25 percent due to early alerts and preventive interventions.
- Patient outcomes improved by 15 percent as a result of continuous monitoring and early issue detection facilitated by IoT and Blockchain integration.
- Example:
  - Diabetic patients received alerts when glucose levels exceeded 180 mg/dL, preventing critical incidents.
  - Hypertensive patients maintained stable blood pressure with real-time notifications and remote monitoring.

9.6 CHALLENGES AND FUTURE DIRECTIONS

9.6.1 Scalability

The scalability limitations that Blockchain faces today prevent it from being able to address the heavy data loads produced by IoT in healthcare.

Challenges

- High latency during high data loads.
- More computational needs for consensus mechanism in large networks.

Solutions

- *Layer 2 scaling solutions*: Secondary networks/solutions, payment channels can accept off-chain transactions to reduce congestion.
- *Sharding*: The division of the Blockchain into smaller segments, or shards, prevents congestion by allowing transactions to occur in parallel, increasing throughput.
- *Edge computing*: Computing at the IoT device can reduce the burden placed on the Blockchain.

9.6.2 Regulatory compliance

The medical care industry involves a lot of stringent laws and regulations with, for example, GDPR in Europe or HIPAA in the United States being two most prominent ones, that are implemented aiming at guarding patient data. A major factor preventing the usage of Blockchain technology is its very nature, that is, it cannot be changed at all which, in most cases, is contradictory to given rules like "the right to be forgotten" as stated in GDPR. There are some encryption-based possibilities for solving the compliance sticking point.

The potential solutions are the following:

1. ZKPs: Provide an opportunity to verify information is valid without seeing the actual information, limiting exposure of the data and maintaining privacy and regulatory compliance.
2. Homomorphic encryption: Allows the ability to conduct analytical operations on data without exposing the underlying sensitive data.
3. Selective data deletion: Newer Blockchain technologies allow for specific record data to be erased without erasing everything.

9.6.3 Interoperability

For the widespread adoption of Blockchain in healthcare to occur, it must fit seamlessly into existing Electronic Health Record (EHR) systems, IoT devices, and healthcare infrastructure.

Challenges

- Limited data-sharing standards.
- The availability of data from existing healthcare systems with the Blockchain.

Solutions

- *APIs for Blockchain–EHR integration*: Standardized APIs can be developed to facilitate data exchange between Blockchain systems and existing EHRs.
- *Development of global standards*: Governments, healthcare providers, and technology companies can work together to develop universal standards of interoperability.
- *Middleware solutions*: Middleware platforms can function as intermediaries between Blockchain and legacy systems, enabling the translation of data.

9.7 CONCLUSION

Blockchain technology is expected to transform healthcare by tackling significant issues like data security, supply chain transparency, and remote patient monitoring. This chapter has illustrated Blockchain's potential to contribute positively toward both patient outcomes and the increased productivity of healthcare organizations through illustrative case studies, like HealthChain and MediLedger.

Key takeaways:

1. Blockchain can securely store sensitive medical data in a highly secure manner.
2. The integration of Blockchain with IoT technology can ensure real-time or constant patient monitoring for the purpose of timely medical intervention if needed.
3. Blockchain advances supply chain transparency aiding in the mitigation of counterfeit drugs and pharmaceutical safety.

Addressing Challenges: While Blockchain technology has great promise, challenges related to scalability, compliance, and interoperability must be addressed. Staying aware of applicable emerging technologies, such as zero-knowledge proofs, sharding, and middleware solutions, can further mitigate obstacles.

Outlook: As Blockchain technology continues to mature, it is expected to be adopted exponentially in healthcare. To leverage the full benefit of Blockchain technology, the healthcare sector will need to support collaboration among stakeholders while enhancing research and development. Together these can evolve a more secure, efficient and patient-centered ecosystem.

REFERENCES

1. Swan, M. (2015). Blockchain: Blueprint for a new economy. *O'Reilly Media.*
2. Azaria, A., Ekblaw, A., Vieira, T., & Lippman, A. (2016). MedRec: Using blockchain for medical data access and permission management. *2nd International Conference on Open and Big Data (OBD)*, 25–30.
3. Ekblaw, A., Azaria, A., Halamka, J.D., & Lippman, A. (2016). A case study on MedRec: Blockchain for medical data sharing. In *Proceedings of the IEEE Conference on Blockchain Technology.*
4. Mettler, M. (2016). Blockchain technology in healthcare: The revolution starts here. *2016 IEEE International Conference on e-Health Networking, Applications and Services (Healthcom)*, 1–3.
5. Yue, X., Wang, H., Jin, D., Li, M., & Jiang, W. (2016). Healthcare data gateways: Found healthcare intelligence on the blockchain with novel privacy risk control. *Journal of Medical Systems*, 40(10), 218.
6. Kshetri, N. (2017). Blockchain's roles in strengthening cybersecurity and protecting privacy. *Telecommunications Policy*, 41(10), 1027–1038.
7. Esposito, C., De Santis, A., Tortora, G., Chang, H., & Choo, K.K.R. (2018). Blockchain: A panacea for healthcare cloud-based data security and privacy? *IEEE Cloud Computing,* 5(1), 31–37.
8. Dwivedi, A.D., Srivastava, G., Dhar, S., & Singh, R. (2019). A decentralized privacy-preserving healthcare Blockchain for IoT. *IEEE Access*, 7, 54088–54100.
9. Angraal, S., Krumholz, H.M., & Schulz, W.L. (2017). Blockchain technology: Applications in healthcare. *Circulation: Cardiovascular Quality and Outcomes*, 10(9), e003800.
10. Gordon, W.J., & Catalini, C. (2018). Blockchain technology for healthcare: Facilitating the transition to patient-driven interoperability. *BMJ Health & Care Informatics*, 25(1), 1–8.
11. World Health Organization (WHO). Estimates related to counterfeit drugs and their implications in healthcare supply chains.
12. Allam, A.H., Gomaa, I., Zayed, H.H., & Taha, M. (2024). IoT-based eHealth using blockchain technology: A survey. *Cluster Computing*, 27, 7083–7110.
13. Zhang, P., Schmidt, D.C., White, J., & Lenz, G. (2018). Blockchain technology use cases in healthcare. *Advances in Computer Science*, 34(3), 59–67.
14. Tandon, A., Dhir, A., Islam, A.N., Mäntymäki, M. (2020). Blockchain in healthcare: A systematic literature review, synthesis, and research agenda. *Computers in Human Behavior*, 104, 106167.
15. Hölbl, M., Kompara, M., Kamišalić, A., & Nemec Zlatolas, L. (2018). A systematic review of the use of blockchain in healthcare. *Symmetry*, 10(10), 470.

Chapter 10

Challenges and future directions for Blockchain of Things (BCoT) in quick commerce

K.V. Haneefa and Harmandeep Singh

10.1 INTRODUCTION

The fast pace of digital innovation has triggered major shifts across industries, driving the rise of new models for operations, strategy, and customer engagement. Two key technologies behind this shift are the Internet of Things (IoT) and Blockchain. IoT refers to networks of connected devices like sensors, machines, and everyday objects that collect, share, and act on data without needing human input (Adhikari & Ramkumar, 2023; De et al., 2022; Khatun et al., 2023; Miaz & Ali, 2020). In parallel, Blockchain is a decentralised and tamper-proof ledger system that records transactions securely across multiple nodes, ensuring transparency and data integrity (Almalki, 2024; Calzadilla & Villa, 2017; Elghaish et al., 2023; Miraz, 2019). When these technologies converge, the result is the Blockchain of Things (BCoT), a framework that enhances operational traceability, security, system compatibility, and overall efficiency (Bhattacharyya et al., 2022; Emam et al., 2020; Miraz, 2020; Yao & Guizani, 2023).

One of the sectors going to benefit most from BCoT is quick commerce (q-commerce), an industry that has rapidly expanded by reshaping last-mile delivery (Haneefa & Singh, 2025a; Sanghi et al., 2023). Q-commerce centres on delivering items like groceries, essentials, and prepared food within extremely short time windows, often between 10 and 60 minutes (Haneefa, 2024; Kapoor et al., 2023; Sanchez, 2024). This model has gained traction as consumers increasingly prioritise speed, convenience, and real-time access to goods. A key feature of q-commerce is the use of "dark stores", small fulfilment hubs placed in urban areas to support fast and local delivery (Kumar & Chidambara, 2023; Rai, 2022; Stojanov, 2022). These facilities are not open to customers; instead, they function as operational sites for order picking, packaging, and dispatching (Rai, 2022). However, managing these stores effectively requires tight coordination across inventory, logistics, and fulfilment, which brings its own challenges like lack of transparency, risk of data manipulation, concerns about product authenticity, and gaps in system security (Kumar & Chidambara, 2023).

DOI: 10.1201/9781003584728-10

BCoT offers a new approach to managing these issues. IoT-enabled devices in dark stores and delivery fleets can collect real-time data on stock levels, temperatures, and order progress. This data, once recorded on a Blockchain, becomes tamper-proof and verifiable (Elghaish et al., 2022). The combination enhances operational oversight, reduces fraud risk, and helps enforce quality standards. It also improves inventory tracking and can lead to smoother operations while supporting higher levels of consumer trust (Jayaprasanna & Sujatha., 2021). Moreover, BCoT can strengthen communication between stakeholders like retailers, suppliers, logistics providers, and customers by allowing data to move securely and efficiently across systems (Yadav & Vishwakarma, 2022).

This study is driven by the rising need for faster and more dependable delivery, and the corresponding stress placed on q-commerce operations. As dark stores become central to meeting this demand, their performance increasingly relies on real-time coordination, accurate stock management, and secure information exchange (Harter et al., 2024). Current systems often fall short of delivering on these fronts. BCoT, with its mix of IoT tracking and Blockchain-backed security, presents an opportunity to close this gap. Yet, its use in q-commerce has not been widely examined. This paper aims to fill that space by analysing how BCoT might improve transparency, support automation, and guide decision-making within dark store logistics. It outlines the key benefits, discusses technical and regulatory hurdles, and suggests practical approaches for adoption. It also identifies areas where further research is needed, especially regarding system integration, scalability, and long-term implications. In doing so, the paper offers a clearer picture of BCoT's relevance to q-commerce and contributes insights for researchers, developers, and industry leaders. The findings reinforce the need for continued inquiry and agile strategies to fully leverage BCoT's potential in building resilient and adaptive q-commerce infrastructures.

10.2 METHODOLOGICAL APPROACH AND FRAMEWORK

The study applies Jaakkola's (2020) typology method to structure the operational, technological, and strategic aspects of BCoT within the q-commerce sector (Haneefa, 2024). This method bridges theoretical concepts with practical challenges, offering a structured lens to explore how BCoT may respond to key industry demands. As noted by Cornelissen (2017), conceptual research plays a crucial role in organising existing knowledge and deepening understanding of emerging fields. The framework centres on essential areas such as real-time inventory tracking, supply chain transparency, system scalability, and collaboration among stakeholders. These focus areas are examined in relation to core BCoT features like distributed ledgers, smart contracts, and secure IoT connectivity. Through this lens, the analysis illustrates how BCoT can address current operational challenges while creating new opportunities for improved efficiency and trust within dark store operations. By drawing together perspectives from Blockchain, IoT, and q-commerce literature, the study offers a coherent model for evaluating BCoT's potential. Though not grounded in empirical data, this conceptual analysis establishes a base for future research and practical adoption, supporting both academic inquiry and strategic implementation.

10.3 THE SIGNIFICANCE OF INTEGRATING BCOT IN DARK STORES

Dark stores, functioning as hyper-local fulfilment centres, have become central to the q-commerce ecosystem, enabling fast deliveries and addressing the rising demand for immediate access to everyday goods (Hwang & Kim, 2023; Schorung, 2023). Strategically located in urban areas and focused on a curated range of high-demand products, these facilities manage inventory, fulfil orders, and coordinate logistics with precision (Kumar & Chidambara, 2023). However, maintaining consistent operational reliability in such settings remains complex, given the involvement of multiple stakeholders like suppliers, retailers, delivery partners, and technology providers often operating with fragmented data systems and limited trust in information exchange (Pache, 2023; Raj et al., 2024).

BCoT provides a decentralised and tamper-resistant foundation that addresses several vulnerabilities present in dark store operations (Rupasena et al., 2021; Yadav & Vishwakarma, 2022). For example, consider a dark store operated by a q-commerce platform in a dense urban area. This store processes hundreds of orders each hour, relying on multiple IoT sensors to monitor perishable goods and track item movement. By integrating BCoT, the platform links all sensor data to a Blockchain ledger, ensuring that temperature breaches, stock changes, and order fulfilment events are recorded in real-time and cannot be altered. When a product recall becomes necessary due to a packaging fault, the system can instantly trace affected items back to a specific batch and supplier, enabling targeted action. This kind of traceability reduces product loss, avoids broad recalls, and reinforces both customer and regulator trust.

Similar integrations have been successfully implemented across various industries: Walmart, in collaboration with IBM, developed a Blockchain-based system to trace food products from farm to store shelf, ensuring food safety and swift responses to contamination issues; Everledger utilises Blockchain to verify the authenticity and ethical sourcing of diamonds, providing transparency in the diamond supply chain; Telstra is experimenting with integrating Blockchain and biometric security to offer enhanced smart home solutions, allowing residents greater control over access and privacy; Power Ledger's platform facilitates peer-to-peer energy trading, using Blockchain to track energy usage and transactions among neighbours;

and Louis Dreyfus Company used Blockchain to track a soybean shipment from the United States to China, improving transparency in the supply chain (Khurana, 2024). By merging the data-capturing capabilities of IoT with the secure, immutable nature of Blockchain, BCoT improves both the transparency and integrity of critical processes. IoT sensors installed in dark stores continuously monitor inventory levels, product conditions, and order timelines (Zubaydi et al., 2023). When this data is recorded on a Blockchain ledger, it becomes immutable and instantly verifiable, allowing for continuous auditing and reducing risks related to inventory miscounts, fraudulent records, or manipulation of operational data (Magrini et al., 2021; Singh & Singh, 2023).

Beyond enhancing security and traceability, BCoT integration supports adaptive operations and informed decision-making within the q-commerce model. With Blockchain-enabled smart contracts, dark store processes such as supplier payments, stock replenishment, or quality verifications can be automated through predefined triggers (Sharma et al., 2024). This automation limits administrative delays, enforces service agreements, and streamlines inventory handling. At the same time, real-time insights from connected devices allow for more responsive stock adjustments, reducing both shortages and surpluses (Yadav & Vishwakarma, 2022).

The value of BCoT adoption also lies in its ability to improve transparency across stakeholder interactions and support compliance with regulatory requirements. Distributed ledger systems can help dark stores demonstrate adherence to safety, ethical sourcing, and sustainability standard factors increasingly important to consumers and regulators alike (Jayaprasanna & Sujatha, 2023; Khatun et al., 2023; Sharma et al., 2024). Enhanced traceability builds consumer confidence in the authenticity and safe handling of products, reinforcing brand reliability. With secure, verifiable data exchange across all actors, BCoT contributes to a more open and coordinated q-commerce ecosystem, enabling collaborative innovation and better use of shared resources across the value chain (Liu et al., 2021; Miaz & Ali, 2020).

10.4 BENEFITS OF BCOT INTEGRATION

The integration of BCoT into dark store operations contributes to a more resilient and responsive ecosystem by supporting key performance areas such as operational efficiency, transparency, scalability, sustainability, regulatory compliance, and customer satisfaction (Yadav & Vishwakarma, 2022). By combining tamper-evident distributed ledger technology with continuous real-time data input from connected devices, BCoT enables a system architecture that is both secure and adaptable. This configuration promotes a more dependable and flexible foundation for q-commerce operations, aligning process integrity with data reliability. In doing so, BCoT has the potential to reshape dark store fulfilment systems into intelligent, traceable, and future-oriented networks capable of handling goods and information with improved precision and accountability.

10.4.1 Operational Efficiency and Transaction Accuracy

The integration of BCoT in dark store operations enables greater automation, precision, and consistency in managing data-driven processes. IoT sensors installed throughout fulfilment centres continuously track essential variables such as stock levels, expiry dates, and temperature conditions (Bhattacharyya et al., 2022). This real-time data is automatically reconciled with Blockchain-based ledgers, ensuring that records such as order verifications and inventory updates remain accurate, timely, and auditable. By reducing manual handling, organisations improve supply chain coordination and lower the risk of human error in tracking inventory (Casino et al., 2019).

These gains in accuracy support more effective demand forecasting and resource distribution. Access to dependable data enables the creation of forecasting models that better anticipate customer needs, helping avoid both excess inventory and stock shortages (Zubaydi et al., 2023). In turn, this improves overall efficiency, limits waste, and cuts down on operational expenses (Adhikari & Ramkumar, 2023). Additionally, access to verified data allows for optimisation of key processes such as dispatching, picking, and packing. This facilitates better labour allocation and more efficient use of delivery vehicles, storage areas, and routing systems.

10.4.2 Enhanced transparency and security

In the context of e-commerce, where consumer trust plays a critical role, BCoT integration offers a mechanism for verifying product authenticity and proper handling across the supply chain. The immutability of Blockchain-based records allows all stakeholders to confirm the source, storage conditions, and delivery schedules of goods, reducing the risks tied to counterfeit products or data tampering (Jayaprasanna & Sujatha, 2023). This transparency helps assure consumers that their purchases are sourced ethically, handled correctly, and stored in line with safety standards, strengthening customer confidence and loyalty (Khatun et al., 2023).

Moreover, Blockchain's cryptographic architecture protects sensitive information from cyber threats. With the growing prevalence of data breaches and unauthorised data manipulation, the security features of Blockchain reduce exposure to malicious activity (Alderazi, 2022). This protection applies to supplier credentials, customer purchase records, and financial transactions preserving their confidentiality, integrity, and authenticity. The traceability enabled by BCoT also enhances quality control practices. In cases requiring product recalls, Blockchain-based records can quickly identify affected batches, suppliers, or storage units, allowing for precise interventions rather than broad, resource-heavy withdrawals (Jayaprasanna & Sujatha, 2023; Liu et al., 2021). This ability to act quickly and accurately helps protect consumer safety while limiting operational and reputational risks.

10.4.3 Scalability and interoperability

The convergence of Blockchain and IoT technologies creates a data-driven framework capable of supporting rapid growth and diversification within q-commerce operations. As expectations for fast delivery increase, dark stores must accommodate larger order volumes, more supplier integrations, and a wider network of logistics providers. A decentralised Blockchain infrastructure enables the seamless integration of heterogeneous IoT devices, systems, and protocols without centralised control (Singh et al., 2019). This structure simplifies the onboarding of new suppliers, couriers, and inventory tools while maintaining secure and consistent data flows.

Interoperability plays a critical role in maintaining technological adaptability. BCoT-enabled platforms are designed to remain compatible with evolving standards, allowing for the incorporation of emerging data types, communication protocols, and device functions (Zhang et al., 2023). This flexibility ensures that dark store ecosystems remain open to technological innovation, avoiding vendor lock-in and supporting long-term scalability. As a result, q-commerce operations benefit from robust and adaptive digital infrastructures that can evolve with shifting market dynamics and customer expectations.

10.4.4 Data-driven decision-making and predictive analytics

A key advantage of integrating Blockchain with IoT in dark store operations lies in its capacity to support more informed, data-driven decisions. Continuous input from sensors, RFID tags, and digital twins feeds real-time data into Blockchain ledgers, offering decision-makers enhanced visibility across the supply chain (Calzadilla & Villa, 2017). These secure and verifiable data streams enable the refinement of delivery routes, inventory control, and scheduling, replacing reliance on estimates or outdated averages with up-to-date operational insights (Khatun et al., 2023).

In addition to improving day-to-day operations, Blockchain-secured IoT data strengthens predictive analytics capabilities. Verified data serves as a reliable foundation for models that anticipate shifts in demand, flag emerging product trends, and detect early signs of disruption across supply nodes (Miraz, 2019). Leveraging this information, machine learning tools can more precisely identify inefficiencies, recommend adjustments, and forecast inventory shortfalls or delivery constraints (Mai et al., 2023). These improvements contribute to greater responsiveness, smoother adaptation to consumer behaviour changes, and more calculated inventory planning, ultimately increasing the agility and competitiveness of dark store operations.

10.4.5 Improved compliance and regulatory adherence

The expansion of e-commerce and q-commerce has brought more stringent regulatory expectations concerning product safety, traceability, and environmental accountability. Integrating Blockchain with IoT technologies offers a structured, resilient framework for ensuring compliance with such requirements. In tightly regulated sectors like pharmaceuticals and perishables, where temperature logs, chain-of-custody records, and batch-level tracking are mandatory, BCoT systems can document these parameters on tamper-proof ledgers, providing verifiable evidence to regulators (Alderazi, 2022).

All actors across the supply chain from upstream suppliers to last-mile carriers operate within a transparent digital environment where compliance can be verified in real time. With permissioned access, auditors and regulators can efficiently review relevant data segments, reducing the need for intrusive checks while maintaining oversight integrity. This streamlined verification process not only simplifies compliance but also lowers the risk of penalties, enhances operational accountability, and reinforces the organisation's standing as a responsible business entity (Calzadilla & Villa, 2017; Khalil et al., 2022).

10.4.6 Cost savings and resource optimisation

Implementing a BCoT ecosystem requires businesses to invest in infrastructure, training, and technology solutions initially, but the long-term financial benefits can be substantial. Automating transactional verifications, inventory reconciliations, and compliance checks significantly reduces labour costs associated with these tasks. Improved forecasting precision and just-in-time inventory practices lower inventory holding costs and minimise waste from expired or unsold goods (Tian, 2017). By enabling more innovative route planning, energy-efficient refrigeration practices, and more targeted resource allocation, BCoT integration reduces transportation and energy expenditures (Yang et al., 2024). Over time, these operational efficiencies and cost savings help businesses offer more competitive pricing to consumers, reinforcing the profitability and sustainability of q-commerce operations.

10.4.7 Sustainability and environmental responsibility

As public awareness of environmental issues grows, organisations increasingly feel pressured to demonstrate sustainability and eco-conscious practices. The traceability offered by BCoT integration helps companies understand and optimise their supply chain's environmental footprint (Liu et al., 2021). By monitoring data such as fuel consumption, route efficiency, packaging material usage, and temperature control conditions, managers identify opportunities to reduce waste and energy consumption (Magrini et al., 2021). For instance, adjusting delivery schedules based on real-time data limits vehicle idling time, while improved inventory management minimises spoilage and reduces unsold stock. Additionally, incorporating sustainable suppliers and partners into the Blockchain-enabled IoT ecosystem ensures that environmental considerations are embedded throughout the supply chain. Verified green credentials, such as eco-friendly packaging or renewable energy use, get recorded immutably on the Blockchain and communicated transparently to consumers. In this way, the integration of BCoT improves operational efficiency and promotes environmental responsibility, positioning q-commerce operations as leaders in responsible business practices.

10.4.8 Enhanced customer experience and loyalty

Businesses leverage customer experience as a key differentiator in an increasingly competitive marketplace (Awini et al., 2023; Haneefa & Singh, 2025b). BCoT integration gives consumers real-time visibility into their orders by allowing them to track delivery progress, verify product authenticity, and confirm compliance with handling standards. This increased transparency and reliability nurtures greater consumer trust and loyalty as customers gain confidence that their purchases are handled with care and delivered as promised (Jayaprasanna & Sujatha., 2021). Additionally, businesses achieve operational efficiencies through BCoT integration, resulting in faster fulfilment times, more accurate order preparation, and fewer missing or incorrect items. These improvements enhance the overall buying experience, encouraging customers to return and share positive feedback with others. Over time, companies that build a reputation for reliable, transparent, and responsible operations can increase their market share and retain customers for the long term (Yadav & Vishwakarma, 2022).

10.4.9 Global standardisation and inter-organisational collaboration

One of the often-overlooked benefits of BCoT integration is its potential to foster global standardisation and collaboration among supply chain participants. By establishing shared, interoperable data standards and communication protocols embedded in Blockchain frameworks, dark stores and their supply chain partners can align their operations more efficiently. This synchronisation leads to smoother operations between suppliers, logistics providers, and delivery partners, ultimately reducing delays, confusion, and data discrepancies (Calzadilla & Villa, 2017; Singh & Singh, 2023; Yadav & Vishwakarma, 2022). As more organisations adopt Blockchain-based IoT solutions, emerging common standards and best practices will simplify the process for new entrants to join established networks. This creates a good cycle where the network becomes increasingly robust, interconnected, and capable of embracing innovative services, technologies, and partnerships.

10.5 CHALLENGES IN INTEGRATING BCOT

Despite the promising advantages of combining Blockchain and IoT technologies within dark store operations, several technical, regulatory, and organisational challenges hinder the effective and widespread integration of BCoT (Akhtar et al., 2021). These issues range from scalability constraints and interoperability hurdles to the nascent nature of emerging standards and ambiguous regulatory landscapes. Understanding these challenges is essential for stakeholders aiming to harness the full potential of BCoT in q-commerce environments.

10.5.1 Technical limitations and scalability concerns

Integrating IoT-enabled sensors with Blockchain networks generates enormous volumes of data, significantly damaging system performance. Traditional Blockchain infrastructures, such as proof-of-work consensus mechanisms, often face latency issues, limited transaction throughput, and elevated energy consumption (Sharma et al., 2024). As dark stores scale their operations to meet increasing customer demands, these performance problems can result in delayed order processing, inefficient inventory updates, and reduced responsiveness to fluctuations in demand (Karunamurthy et al., 2023). Even more energy-efficient consensus algorithms, such as proof-of-stake, may struggle to handle large-scale data influxes and real-time processing requirements if not optimised for IoT-centric applications (Pattnayak & Patnaik, 2022).

10.5.2 Interoperability and standardisation gaps

Different IoT platforms, communication protocols, and data formats create challenges for building a seamless Blockchain-based ecosystem for q-commerce. Supply chain partners struggle to integrate their systems into a unified ledger without widely adopted standards for data exchange, cryptographic protocols, and device authentication (Adhikari & Ramkumar, 2023; Dai et al., 2019). Inconsistent data models and incompatible application programming interfaces (APIs) restrict the smooth flow of information, undermining the transparency and trust that Blockchain aims to provide.

To address these interoperability gaps, stakeholders must develop middleware (Software acts as a bridge between applications, systems, or databases, enabling communication and data management through common services) solutions, data conversion tools, and multistakeholder groupings focused on establishing common frameworks for BCoT integration (Fremantle & Scott, 2017; Leng et al., 2022).

10.5.3 Regulatory uncertainty and compliance

While Blockchain's decentralised structure offers potential data control and trust benefits, it complicates compliance with regional data protection and consumer safety regulations. National and international laws governing data privacy, financial transactions, and product authenticity vary significantly (Casino et al., 2019). Organisations deploying BCoT must ensure they conform to relevant legal frameworks, which can be challenging, as immutable ledgers may conflict with regulations like the General Data Protection Regulation (GDPR). This regulation grants individuals rights to data removal and modification (Finck, 2018). Additionally, governing bodies often struggle to classify Blockchain-based assets and transactions, leaving stakeholders uncertain about taxation, auditing requirements, and liability issues. This regulatory ambiguity can discourage investment and slow the pace of innovation in BCoT solutions.

10.5.4 Organisational resistance and cultural barriers

Adopting Blockchain technology in dark stores and their associated supply chains requires firms to fundamentally change their operational routines, decision-making processes, and organisational structures (Lo et al., 2019). Companies that rely on traditional data management systems and hierarchical command structures often resist embracing Blockchain's decentralised ethos and increased transparency. Internal stakeholders may perceive these changes as threats to authority, intellectual property, or established vendor relationships. Companies need to implement extensive training, align incentives, and develop strong change management initiatives to overcome this organisational resistance (Elghaish et al., 2021).

10.5.5 Nascent state of BCoT research and speculative benefits

While researchers have documented the theoretical advantages of BCoT, practical research and large-scale testbeds are still limited. Many Blockchain-based IoT use cases remain in pilot stages or controlled laboratory settings, which makes it challenging to predict long-term benefits or assess the return on investment reliably (Casino et al., 2019; Miraz, 2020). This developing stage breeds uncertainty and speculation, causing q-commerce stakeholders to hesitate in committing resources to technologies that may have unproven and potentially exaggerated value.

10.5.6 Network latency and data throughput issues

In BCoT, high-speed data transmission is critical, especially for time-sensitive operations like inventory tracking, order processing, and supply chain logistics. However, even when optimised for IoT, Blockchain networks often face challenges with network latency and data quantity limitations (Lo et al., 2019). The time it takes to spread and validate blocks across the network hinders data transfer speed between IoT devices and the Blockchain. For large-scale dark store operations, where thousands of devices transmit data in real-time, this latency creates significant problems (Gupta & Singh, 2022; Naidu et al., 2018; Rizwan et al., 2022).

Network congestion becomes serious when transactions accumulate and require validation before entering the Blockchain. As the number of transactions increases, validators need more time to process each one, which leads to delays in processing orders, updating inventories, or making real-time decisions using data from IoT devices. Furthermore, when Blockchain transactions are not processed in real time, discrepancies can arise in the recorded data, undermining customer trust, complicating inventory management, and compromising the system's overall reliability (Karunamurthy et al., 2023; Raza & Singh, 2022). To tackle these issues, Blockchain networks need to evolve and effectively manage the high volumes demanded by large-scale IoT applications. Implementing technologies like sharding (a method of partitioning a Blockchain into smaller pieces for parallel processing to improve scalability) or layer 2 (secondary protocols built on top of a Blockchain to enhance speed and reduce costs), solutions will enable parallel transaction processing, which will help reduce network congestion and latency, allowing BCoT to scale efficiently (Leng et al., 2022; Liu et al., 2023; Yu et al., 2020).

10.5.7 Energy consumption and environmental impact

The energy consumption associated with Blockchain technologies, particularly those using Proof-of-Work consensus mechanisms, sparks ongoing debate. As the industry witnesses the rise of Proof-of-Stake (PoS) and other energy-efficient consensus algorithms, the overall environmental impact of Blockchain continues to raise concerns. In dark stores, IoT systems rely on multiple devices that constantly transmit data to the Blockchain, creating a constant energy demand (Elghaish et al., 2023; Sharma et al., 2024). The growing deployment of Blockchain networks to manage these devices risks generating a significant environmental footprint. For instance, with thousands of IoT devices consistently feeding data into the Blockchain, their combined energy consumption can escalate rapidly when using PoW-based Blockchains (Miraz et al., 2020). This challenge poses a risk for companies striving to meet sustainability goals or comply with environmental regulations. Subsequently, discovering energy-efficient Blockchain solutions tailored to IoT applications becomes a pressing issue. To facilitate BCoT integration, we must prioritise sustainable energy consumption practices and adopt carbon-neutral Blockchain protocols. Some Blockchain networks are already investigating hybrid approaches to consensus mechanisms that lower energy requirements, but these solutions remain in the early stages and necessitate further development and testing (Elghaish et al., 2023).

10.5.8 Integration with legacy systems

In many dark store operations, legacy systems such as older inventory management software, point-of-sale (POS) systems, and supply chain management tools remain in use. Integrating Blockchain and IoT technologies with these legacy systems can be complex and costly (Jandl et al., 2019; Khalil et al., 2022). Many of these legacy systems were not designed with Blockchain or IoT in mind and lacked the compatibility to interface with modern Blockchain infrastructures seamlessly. Transitioning to a Blockchain-based system demands a complete replacement or a substantial upgrade of legacy systems. This process can be resource-intensive, both financially and in terms of time, as organisations undertake data migration, system reengineering, and training for staff accustomed to old operational methods (Calzadilla & Villa, 2017). Even after upgrading a legacy system, companies may face integration challenges that lead to inefficiencies, errors, or data loss during the transition phase. Moreover, companies often hesitate to replace or overhaul existing systems due to concerns about potential downtime, disruptions to operations, or the risk of system failure (Tedeschi et al., 2018). Therefore, significant planning and investment are essential to ensure a smooth integration of Blockchain and IoT with current infrastructure.

10.6 POTENTIAL SOLUTIONS AND FUTURE RESEARCH DIRECTIONS

To address the challenges of integrating BCoT in dark stores, we need a multidisciplinary approach that includes technological innovation, policy reform, organisational change management, and ongoing research initiatives. We should focus on building more efficient Blockchain architectures, fostering interoperability, clarifying regulatory frameworks, and cultivating an organisational culture that embraces decentralised, data-driven operations.

10.6.1 Advanced consensus mechanisms and scalable architectures

Researchers and developers are actively exploring next-generation consensus mechanisms that can mitigate the performance constraints of current Blockchain networks. They investigate algorithms like Practical Byzantine Fault Tolerance (PBFT), Delegated Proof-of-Stake (DPoS), and Directed Acyclic Graph (DAG)-based ledgers (Bano et al., 2019) to handle the vast data volume from IoT devices in dark stores (Lo et al., 2019). Layer-two solutions, such as payment channels and sidechains, can enhance scalability by offloading some transactions from the main Blockchain (Singh et al., 2022;Yang et al., 2022). Future work will delve into the trade-offs between throughput, security, and energy efficiency in these innovative architectures, tailoring protocols to the unique data environments of q-commerce operations.

10.6.2 Interoperability frameworks and standardisation initiatives

Standardised communication protocols must be developed and adopted to achieve seamless data exchange among diverse IoT platforms and Blockchain networks. Industry consortia, such as the IEEE P2418 working groups, are laying the groundwork for compatible frameworks that define common data structures, security requirements, and identity management tools for IoT and Blockchain integration (IEEE, 2022). Beyond formal standards, open-source projects and middleware solutions can help simplify complications and provide vendors with toolkits for integrating disconnected systems (Belchior et al., 2020). Further research can evaluate the success of these interoperability efforts by examining their impact on supply chain coordination, data accuracy, and stakeholder trust.

10.6.3 Regulatory clarity and compliance mechanisms

Consolidating a global regulatory approach to Blockchain and IoT data governance remains a crucial long-term goal. Policymakers and industry stakeholders should collaborate to create guidelines that acknowledge Blockchain's immutable nature while respecting individuals' rights to data privacy and rectification (Finck, 2018). We can incorporate mechanisms like zero-knowledge proofs and selective disclosure into Blockchain designs, allowing compliance with privacy regulations without undermining transparency (Dai et al., 2019). Empirical research can focus on evaluating how these privacy-preserving techniques align with regional regulations such as the GDPR, the California Consumer Privacy Act (CCPA), or emerging international standards.

10.6.4 Organisational change management and training

Organisations need structured change management strategies to overcome internal resistance and foster an innovative culture. Investing in employee training, workshops, and educational resources will help staff understand Blockchain fundamentals, understand the significance of decentralised data management, and recognise the strategic value of BCoT-enabled transparency (Cai et al., 2021; Lo et al., 2019). Research can explore the effectiveness of incentive structures encouraging cross-functional collaboration and knowledge exchange. Developing best practices for implementing Blockchain projects, such as pilot programmes and phased rollouts, and establishing centres of excellence can guide firms in seamlessly integrating BCoT into their workflows.

10.6.5 Empirical validation and pilot projects

We require more empirical, long-term studies to validate the expected benefits of BCoT in q-commerce. Conducting pilot projects in controlled environments, followed by scaled implementations in operational dark stores, will generate data to assess performance metrics such as order processing speed, inventory accuracy, and energy efficiency (Queiroz & Fosso Wamba, 2019). Comparative studies that examine conventional supply chains alongside BCoT-enhanced ones can provide quantifiable evidence for return on investment. Research can also investigate how integrating complementary emerging technologies, such as edge computing, artificial intelligence, and 6G connectivity, can strengthen the impact of BCoT. These combined insights will enhance theoretical frameworks, guide technology roadmaps, and establish best practice guidelines as the market matures.

10.7 CONCLUSION

Integrating Blockchain of Things (BCoT) in the q-commerce sector, particularly within dark store operations, represents a significant shift with major implications for operational efficiency, transparency, and trust. As demand for faster fulfilment continues to grow, businesses must adapt by embedding systems that ensure precision, reliability, and alignment across actors. This chapter outlined the ways BCoT contributes to addressing core operational constraints in q-commerce and enables firms to respond to consumer expectations more effectively. Dark stores operate as decentralised fulfilment hubs that rely on close coordination between retailers, suppliers, and logistics partners.

In this context, BCoT creates a shared, tamper-resistant record of all transactions. By making these records accessible to relevant stakeholders, the system helps reduce uncertainty and builds trust. Issues such as product tracking, stock availability, and order accuracy can be handled more reliably when transaction data is drawn from real-time sources linked through IoT devices. The distributed nature of BCoT also offers scalability, allowing firms to connect additional systems, tools, and partners without major overhauls. This makes it easier to expand to new locations, incorporate new data sources, and support wider collaboration. With secure, real-time information sharing, firms are better positioned to manage change and make informed operational choices. These capabilities are especially relevant in fast-moving sectors like q-commerce, where margin and time pressures are constant.

However, the implementation of BCoT faces practical hurdles. Blockchain's ability to process and scale with high-volume IoT data remains a concern. The lack of uniform standards in IoT device protocols further complicates system integration. Regulatory uncertainty around data use, ownership, and privacy also introduces risk. Without a clear regulatory framework, firms may hesitate to invest in full-scale adoption. To address these issues, future research must focus on improving system scalability and creating consistent interoperability frameworks for connected devices. Establishing shared technical and regulatory standards will require coordinated efforts across firms, regulators, and academic researchers. Applied studies on how BCoT performs in live q-commerce environments can generate insights that inform both policy and practice. The potential benefits of BCoT for dark store operations such as more reliable data, stronger coordination, and improved fulfilment can only be realised if these barriers are addressed. Cross-sector collaboration is essential to move from experimental use to broad implementation. If executed thoughtfully, BCoT can support leaner, faster, and more accountable systems in q-commerce. In doing so, it helps businesses remain responsive to changing consumer needs while improving how inventory, transactions, and service flows are managed at scale.

DATA AVAILABILITY STATEMENT

Data sharing is not applicable to this book/chapter as no new data were created or analysed in this study.

REFERENCES

Adhikari, N., & Ramkumar, M. (2023). IoT and Blockchain Integration: Applications, Opportunities, and Challenges. *Network*, *3*(1), 115–141. https://doi.org/10.3390/network3010006

Akhtar, P., Azima, N., Ghafar, A., & Din, S. U. (2021). Barricades in the Adoption of Block-Chain Technology in Supply Chain Management: Challenges and Benefits. *Transnational Marketing Journal*, *9*(1). https://doi.org/10.33182/tmj.v9i1.1021

Alderazi, F. (2022). Security of Internet of Things: A Review of Challenges with Integrating Blockchain with IoT. In *Proceedings of 2022 2nd International Conference on Computing and Information Technology, ICCIT 2022*. https://doi.org/10.1109/ICCIT52419.2022.9711650

Almalki, J. (2024). State-of-the-Art Research in Blockchain of Things for HealthCare. *Arabian Journal for Science and Engineering*, *49*(3), 3163–3191. https://doi.org/10.1007/s13369-023-07896-5

Awini, G., Alolo Akeji, A. A. R., Majeed, M., & Mensah, K. (2023). Does Blockchain and Internet of Things (BCoT) Help in Managing Customer Data? In *The Rise of Blockchain Applications in Customer Experience*, edited by Mohammed Majeed, Kwame Simpe Ofori, George Kofi Amoako, Abdul-Raheed Alolo, and Gideon Awini. IGI Global Scientific Publication. https://doi.org/10.4018/978-1-6684-7649-9.ch002

Bano, S., Sonnino, A., Al-Bassam, M., Azouvi, S., McCorry, P., Meiklejohn, S., & Danezis, G. (2019). SoK: Consensus in the Age of Blockchains. In *Proceedings of the 1st ACM Conference on Advances in Financial Technologies*, 183–198. https://doi.org/10.1145/3318041.3355458

Belchior, R., Vasconcelos, A., Guerreiro, S., & Correia, M. (2020). A Survey on Blockchain Interoperability: Past, Present, and Future Trends. *ACM Computing Surveys*, *54*. https://doi.org/10.1145/3471140

Bhattacharyya, S., Banerjee, P. S., Karmakar, A., De, D., & Rodrigues, J. J. P. C. (2022). BCoT: Concluding Remarks. In *Lecture Notes on Data Engineering and Communications Technologies,* edited by Debashis De, Siddhartha Bhattacharyya, and Joel J. P. C. Rodrigues. Springer Nature. https://doi.org/10.1007/978-981-16-9260-4_12

Cai, G., Xuran, L., Yanhua, C., & Hongning, D. (2021). Blockchain Technology for Internet of Things: An Overview. *Chinese Journal on Internet of Things*, *5*. https://doi.org/10.11959/j.issn.2096-3750.2021.00201

Calzadilla, J. F., & Villa, A. (2017). Systematic Literature Review of the Use of Blockchain in Supply Chain. *12th European Research Seminar (ERS) on Logistics and SCM*. http://oa.upm.es/51171/1/INVE_MEM_2017_278222.pdf

Casino, F., Dasaklis, T. K., & Patsakis, C. (2019). A Systematic Literature Review of BBlockchain-Based Applications: Current Status, Classification and Open Issues. *Telematics and Informatics*, *36*, 55–81. https://doi.org/10.1016/j.tele.2018.11.006

Cornelissen, J. (2017). Editor's Comments: Developing Propositions, a Process Model, or a Typology? Addressing the Challenges of Writing Theory Without a Boilerplate. *Academy of Management Review*, *42*(1), 1–9. https://doi.org/10.5465/amr.2016.0196

Dai, H. N., Zheng, Z., & Zhang, Y. (2019). Blockchain for Internet of Things: A Survey. *IEEE Internet of Things Journal*, *6*(5), 8076–8094. https://doi.org/10.1109/JIOT.2019.2920987

De, D., Karmakar, A., Banerjee, P. S., Bhattacharyya, S., & Rodrigues, J. J. P. C. (2022). BCoT: Introduction to Blockchain-Based Internet of Things for Industry 5.0. In *Lecture Notes on Data Engineering and Communications Technologies,* edited by Debashis De, Siddhartha Bhattacharyya, and Joel J. P. C. Rodrigues. Springer Nature. https://doi.org/10.1007/978-981-16-9260-4_1

Elghaish, F., Hosseini, M. R., Matarneh, S., Talebi, S., Wu, S., Martek, I., Poshdar, M., & Ghodrati, N. (2021). Blockchain and the 'Internet of Things" for the Construction Industry: Research Trends and Opportunities.' *Automation in Construction*, *132*, 103942. https://doi.org/10.1016/j.autcon.2021.103942

Elghaish, F., Pour Rahimian, F., Brooks, T., Dawood, N., & Abrishami, S. (2023). Blockchain of Things and Deep Learning Applications in Construction. In *Blockchain of Things and Deep Learning Applications in Construction: Digital Construction Transformation*, edited by Faris Elghaish, Farzad Pour Rahimian, Tara Brooks, Nashwan Dawood, and Sepehr Abrishami. Springer International Publishing. https://doi.org/10.1007/978-3-031-06829-4

Elghaish, F., Rahimian, F. P., & Dawood, N. (2022). *Decentralized Maintenance Supply Chain System for Highly Sensitive Assets: 'Blockchain of Things' (BCoT)-Based Solution*. https://doi.org/10.3311/ccc2022-005

Emam, O., Fahmy, H., & Mamdouh, M. (2020). Securing IoT Systems Using Blockchain Algorithms. *Communications on Applied Electronics, 7*. https://doi.org/10.5120/cae2020652871

Finck, M. (2018). Blockchains and the General Data Protection Regulation. In *Blockchain Regulation and Governance in Europe* (pp. 88–116). Cambridge University Press. https://doi.org/10.1017/9781108609708.004

Fremantle, P., & Scott, P. (2017). A Survey of Secure Middleware for the Internet of Things. *PeerJ Computer Science*, *3*, e114. https://doi.org/10.7717/peerj-cs.114

Gupta, A., & Singh, S. (2022). Application of Industry 4.0 Technologies in Sustaining Supply Chain. *2022 3rd International Conference on Intelligent Engineering and Management (*ICIEM*)*, 396–402. https://doi.org/10.1109/ICIEM54221.2022.9853157

Haneefa, K. V. (2024). Quick Commerce in the Quantum Age. In *The Quantum AI Era of Neuromarketing,* edited by Seda Yıldırım, Işıl Demirtaş, and Firdous Ahmad Malik. IGI Global Scientific Publishing, pp. 403–428. https://doi.org/10.4018/979-8-3693-7673-7.ch017

Haneefa, K. V., & Singh, H. (2025a). From Convenience to Sustainability. In *Green Economics and Strategies for Business Sustainability,* edited by Rhytheema Dulloo, Anand Kurian, Minja Bolesnikov, Ilse Struweg, and Kaliyan Mathiyazhagan. IGI Global (pp. 261–288). https://doi.org/10.4018/979-8-3693-8949-2.ch011

Haneefa, K. V., & Singh, H. (2025b). Enhancing Quick Commerce Service Experience Through AI Marketing. In *Leveraging AI-Powered Marketing in the Experience-Driven Economy* (Vol. 1, pp. 203–228). IGI Global. https://doi.org/10.4018/979-8-3693-9561-5.ch007

Harter, A., Stich, L., & Spann, M. (2024). The Effect of Delivery Time on Repurchase Behavior in Quick Commerce. *Journal of Service Research, 28*. https://doi.org/10.1177/10946705241236961

Hwang, J., & Kim, Y. (2023). Challenges of Dark Stores and Urban Planning: A Case Study of B-Mart, Seoul. *Preprints*, May, *1,* 1–16. https://doi.org/10.20944/preprints202305.0084.v1

IEEE. (2022). *IEEE SA—The IEEE Standards Association*. The IEEE Standards Association. https://standards.ieee.org/

Jaakkola, E. (2020). Designing Conceptual Articles: Four Approaches. *AMS Review*, *10*(1–2), 18–26. https://doi.org/10.1007/s13162-020-00161-0

Jandl, C., Nurgazina, J., Schoffer, L., Reichl, C., Wagner, M., & Moser, T. (2019). SensiTrack—A Privacy by Design Concept for Industrial IoT Applications. *2019 24th IEEE International Conference on Emerging Technologies and Factory Automation (ETFA)*, 1782–1789. https://doi.org/10.1109/ETFA.2019.8869186

Jayaprasanna, M. C., & Sujatha, S. (2023). The Integration of Blockchain with Internet-of-Things (BCOT) in Smart cities. In *Blockchain Enabled Secure Big Data Computing for Smart Cities Using Internet of Things*, edited by Ashok Kumar, D. Karthika Renuka, V. Chandrasekar, and Sonali Agarwal. Nova Science Publisher.

Kapoor, A., Sindwani, R., & Goel, M. (2023). Exploring Quick Commerce Service Experience: A Moderated Mediated Investigation Using SEM and fsQCA. *Total Quality Management and Business Excellence, May, 34*, 1–24. https://doi.org/10.1080/14783363.2023.2213653

Karunamurthy, D. A., Amalraj Victoire, D. T., Sunali, S., & Thamizharasi, P. (2023). Block-Chain Management in Supply Chain Management—A Comprehensive Review. *Quing: International Journal of Innovative Research in Science and Engineering*, *2*(2), 50–77. https://doi.org/10.54368/qijirse.2.2.0006

Khalil, U., Mueen-Uddin, Malik, O. A., & Hussain, S. (2022). A Blockchain Footprint for Authentication of IoT-Enabled Smart Devices in Smart Cities: State-of-the-Art Advancements, Challenges and Future Research Directions. *IEEE Access*, *10*, 76805–76823. https://doi.org/10.1109/ACCESS.2022.3189998

Khatun, M. S., Liu, A., & Miraz, M. H. (2023). BCoT-Based Smart Manufacturing: An Enhanced Precise Measurement Management System. In *Lecture Notes of the Institute for Computer Sciences, Social-Informatics and Telecommunications Engineering, LNICST,* edited by Mahdi H. Miraz, Garfield Southall, Maaruf Ali, and Andrew Ware. Springer. https://doi.org/10.1007/978-3-031-25161-0_3

Khurana, A. (2024). *How Blockchain and IoT Revolutionize Smart Cities*. Oodles Blockchain. https://Blockchain.oodles.io/blog/Blockchain-and-iot-revolutionize-smart-cities-real-world-examples/

Kumar, I., & Chidambara. (2023). Freight Patterns and Spatial Planning Requirements of Third Generation E-Commerce in Indian Cities. *Urban, Planning and Transport Research*, *11*(1). https://doi.org/10.1080/21650020.2023.2256817

Leng, J., Chen, Z., Huang, Z., Zhu, X., Su, H., Lin, Z., & Zhang, D. (2022). Secure Blockchain Middleware for Decentralized IIoT towards Industry 5.0: A Review of Architecture, Enablers, Challenges, and Directions. *Machines*, *10*(10), 858. https://doi.org/10.3390/machines10100858

Liu, A., Khatun, M. S., Liu, H., & Miraz, M. H. (2021). Lightweight Blockchain of Things (BCoT) Architecture for Enhanced Security: A Literature Review. In *Proceedings – 2021 International Conference on Computing, Networking, Telecommunications and Engineering Sciences Applications, CoNTESA 2021*. https://doi.org/10.1109/CoNTESA52813.2021.9657112

Liu, X., Xie, H., Yan, Z., & Liang, X. (2023). A Survey on BBlockchain Sharding. *ISA Transactions*, *141*, 30–43. https://doi.org/10.1016/j.isatra.2023.06.029

Lo, S. K., Liu, Y., Chia, S. Y., Xu, X., Lu, Q., Zhu, L., & Ning, H. (2019). Analysis of Blockchain Solutions for IoT: A Systematic Literature Review. *IEEE Access*, *7*, 58822–58835. https://doi.org/10.1109/ACCESS.2019.2914675

M.C., Jayaprasanna, & S, Sujatha. (2021). A Block-Chain of Things (BcoT) Based System for Detecting Counterfeit Products in Supply Chain Management. *SSRN Electronic Journal.* https://doi.org/10.2139/ssrn.3852852

Magrini, C., Nicolas, J., Berg, H., Bellini, A., Paolini, E., Vincenti, N., Campadello, L., & Bonoli, A. (2021). Using Internet of Things and Distributed Ledger Technology for Digital Circular Economy Enablement: The Case of Electronic Equipment. *Sustainability (Switzerland), 13.* https://doi.org/10.3390/su13094982

Mai, T., Yao, H., Zhang, N., Xu, L., Guizani, M., & Guo, S. (2023). Cloud Mining Pool Aided Blockchain-Enabled Internet of Things: An Evolutionary Game Approach. *IEEE Transactions on Cloud Computing, 11.* https://doi.org/10.1109/TCC.2021.3110965

Miaz, M. H., & Ali, M. (2020). Integration of BBlockchain and IoT: An Enhanced Security Perspective. *Annals of Emerging Technologies in Computing, 4.* https://doi.org/10.33166/AETiC.2020.04.006

Miraz, M. H. (2019). Blockchain of Things (BCoT): The Fusion of Blockchain and IoT Technologies. *SSRN Electronic Journal, 60.* https://doi.org/10.2139/ssrn.3464085

Miraz, M. H. (2020). Blockchain of Things (BCoT): The Fusion of Blockchain and IoT Technologies. In *Studies in Big Data,* edited by Mahdi H. Miraz. Springer. https://doi.org/10.1007/978-981-13-8775-3_7

Miraz, M. H., Ali, M., Miaz, M. H., & Ali, M. (2020). Integration of Blockchain and IoT: An Enhanced Security Perspective. *Annals of Emerging Technologies in Computing*, *4*(4), 52–63. https://doi.org/10.33166/AETiC.2020.04.006

Naidu, V., Mudliar, K., Naik, A., & Bhavathankar, P. (2018). A Fully Observable Supply Chain Management System Using Block Chain and IOT. *2018 3rd International Conference for Convergence in Technology (I2CT)*, 1–4. https://doi.org/10.1109/I2CT.2018.8529725

Pache, G. (2023). Q-commerce Logistical Networks: A Shift in Digital Retail Towards "Going Dark"? *EMNet 2023, 1.*

Pattnayak, P., & Patnaik, S. (2022). Beauty of Block Chain Technology in Supply Chain Management. *Trends in Sciences*, *19*(10), 4175. https://doi.org/10.48048/tis.2022.4175

Queiroz, M. M., & Fosso Wamba, S. (2019). Blockchain Adoption Challenges in Supply Chain: An Empirical Investigation of the Main Drivers in India and the USA. *International Journal of Information Management*, *46*, 70–82. https://doi.org/10.1016/j.ijinfomgt.2018.11.021

Rai, H. B. (2022). Dark Stores as Post-pandemic Omnichannel Strategy: Implications for Urban Logistics (forthcoming). In *Routledge Handbook of Urban Logistics,* edited by Jason Monios, Lucy Budd, and Stephen Ison. Routledge, pp. 134–146. https://doi.org/10.4324/9781003241478-13

Raj, G., Roy, D., de Koster, R., & Bansal, V. (2024). Stochastic Modeling of Integrated Order Fulfillment Processes with Delivery Time Promise: Order Picking, Batching, and Last-mile Delivery. *European Journal of Operational Research, 316.* https://doi.org/10.1016/j.ejor.2024.03.003

Raza, Z., & Singh, A. K. (2022). A Framework for the Blockchain and IoT-Based Supply Chain Management System. *International Journal of Applied Logistics*, *12*(1), 1–30. https://doi.org/10.4018/IJAL.309090

Rizwan, A., Karras, D. A., Kumar, J., Sánchez-Chero, M., Mogollón Taboada, M. M., & Altamirano, G. C. (2022). An Internet of Things (IoT) Based Block Chain Technology to Enhance the Quality of Supply Chain Management (SCM). *Mathematical Problems in Engineering, 2022*, 1–12. https://doi.org/10.1155/2022/9679050

Rupasena, J., Rewa, T., Hemachandra, K. T., & Liyanage, M. (2021). Scalable Storage Scheme for BBlockchain-Enabled IoT Equipped Food Supply Chains. *2021 Joint European Conference on Networks and Communications and 6G Summit, EuCNC/6G Summit 2021*, 300–305. https://doi.org/10.1109/EuCNC/6GSummit51104.2021.9482449

Sanchez, P. E. L. (2024). *An Analysis of the Drivers of Consumers' Purchasing Behavior in Quick Commerce Platforms*. https://aaltodoc.aalto.fi/handle/123456789/127048

Sanghi, N., Chandra Balodi, K., & Gupta, V. (2023). The Emergence of the Indian Hyperlocal Grocery Delivery Industry: Dunzo v/s Blinkit. *Journal of Information Technology Teaching Cases, 14*. https://doi.org/10.1177/20438869231151449

Schorung, M. (2023). *Quick Commerce: Will the Disruption of the Food Retail Industry Happen? Investigating the Quick Commerce Supply Chain and the Impacts of Dark Stores*. https://shs.hal.science/halshs-04098596

Sharma, S., Sharma, T., Tiwari, A., & Gupta, S. (2024). Streamlining IoT-driven Data Using Blockchain. *International Research Journal on Advanced Engineering and Management (IRJAEM)*, 2(05), 1509–1514. https://doi.org/10.47392/irjaem.2024.0204

Singh, A., Kumar, G., Saha, R., Conti, M., Alazab, M., & Thomas, R. (2022). A Survey and Taxonomy of Consensus Protocols for BBlockchains. *Journal of Systems Architecture*, *127*, 102503. https://doi.org/10.1016/j.sysarc.2022.102503

Singh, A., Shujaat, G., Singh, I., Tripathi, A., & Thakur, D. (2019). A Survey of Blockchain Technology Security. *International Journal of Computer Engineering in Research Trends, 6*. https://doi.org/10.22362/ijcert/2019/v6/i04/v6i0401

Singh, I., & Singh, B. (2023). Integration of Decentralized Blockchain with Cloud & IoT Based SCM. *2023 International Conference on Advancement in Computation and Computer Technologies, InCACCT 2023*. https://doi.org/10.1109/InCACCT57535.2023.10141797

Stojanov, M. (2022). *Q-Commerce – The Next Generation*. http://hdl.handle.net/10610/4683

Tian, F. (2017). A Supply Chain Traceability System for Food Safety Based on HACCP, Blockchain & Internet of Things. *International Conference on Service Systems and Service Management, Dalian*, 1–6.

Tedeschi, S., Rodrigues, D., Emmanouilidis, C., Erkoyuncu, J., Roy, R., & Starr, A. (2018). A Cost Estimation Approach for IoT Modular Architectures Implementation in Legacy Systems. *Procedia Manufacturing*, *19*, 103–110. https://doi.org/10.1016/j.promfg.2018.01.015

Yadav, A. K., & Vishwakarma, V. P. (2022). Adoptation of Blockchain of Things(BCOT): Oppurtunities & Challenges. *2022 IEEE International Conference on Blockchain and Distributed Systems Security, ICBDS 2022*. https://doi.org/10.1109/ICBDS53701.2022.9935985

Yang, L., Zou, Y., Xu, M., Xu, Y., Yu, D., & Cheng, X. (2022). Distributed Consensus for Blockchains in Internet-of-Things Networks. *Tsinghua Science and Technology*, *27*(5), 817–831. https://doi.org/10.26599/TST.2021.9010065

Yang, X., Ostermeier, M., & Hübner, A. (2024). Winning the Race to Customers with Micro-Fulfillment Centers: An Approach for Network Planning in Quick Commerce. *Central European Journal of Operations Research, 32*. https://doi.org/10.1007/s10100-023-00893-x

Yao, H., & Guizani, M. (2023). Blockchain-Enabled Intelligent IoT. In *Wireless Networks (United Kingdom),* edited by Haipeng Yao and Mohsen Guizani. Springer. https://doi.org/10.1007/978-3-031-26987-5_7

Yu, G., Wang, X., Yu, K., Ni, W., Zhang, J. A., & Liu, R. P. (2020). Survey: Sharding in Blockchains. *IEEE Access*, *8*, 14155–14181. https://doi.org/10.1109/ACCESS.2020.2965147

Zhang, M., Yang, M., Shen, G., Xia, Z., & Wang, Y. (2023). A Verifiable and Privacy-Preserving Cloud Mining Pool Selection Scheme in BBlockchain of Things. *Information Sciences, 623*. https://doi.org/10.1016/j.ins.2022.11.169

Zubaydi, H. D., Varga, P., & Molnár, S. (2023). Leveraging Blockchain Technology for Ensuring Security and Privacy Aspects in Internet of Things: A Systematic Literature Review. In *Sensors,* edited by Raffaele Bruno et al. Multidisciplinary Digital Publishing Institute, Vol. 23, Issue 2, p. 788. https://doi.org/10.3390/s23020788

Chapter 11

Research challenges and future trends in Blockchain of Things

Shraddha C. Subhedar and Deepa Parasar

With the rapid development and evolution of modern information and communication technologies, Industry 4.0 has upturned the conventional industrial development model. Fourth Industrial Revolution, i.e., Industry 4.0, implies the digital transformation of not only manufacturing sectors such as the automotive industry but it also brought transition to other sectors such as chemical, retail, mining, logistics, power generation, oil, construction, and gas industries with incorporation of AI and automation. Industry 4.0 brings together cutting-edge technologies like big data, artificial intelligence (AI), the Internet of Things (IoT), and blockchain technology. The feature of self-configuration capability is a main reason behind IoT to be a dynamically growing global infrastructure. Only the necessary condition for this is that this configuration must conform to standard and interoperable communication protocols. These arrangements allow all components to interface and converse with each other by enabling information sharing and collaborative decision-making. So, we can say that the inter-network where several elements—infrastructures, vehicles, smart physical devices—are interfaced with each other and these elements are implanted with software, electronic components, actuators, sensors, and network connections to enable them for easy and quick data collection and data exchange, is known as Internet of Things [1]. We are moving toward a more futuristic world where everything is connected and done over the internet. Blockchain of Things (BCoT) refers to the alliance of two thriving technologies: Blockchain and Internet of Things (IoT). International Data Corporation calculated that 41 billion IoT devices connected to the internet will be generating 79.4 zettabytes of data by 2026. Although IoT has given us new ways to connect with our surroundings, it incurs concerns about privacy, security, and cost as it exploits functionality of centralized architecture. The decentralization, distributed nature, and immutability of Blockchain technology make it a competent technology for IoT as these crucial features of BC can address issues related to trust and security, high maintenance cost, and transparency problems, and so on. Let's delve into the basic concepts that need to be understood to get better insights of BCoT.

 DOI: 10.1201/9781003584728-11

11.1 IOT + BLOCKCHAIN => BCOT

The network of physical devices such as home appliances, machines, and various digital objects referred to as "things", is called Internet of Things (IoT). If we consider the term "*thing*", definition wise and purpose-based, then we can say that it's a very generic in nature. It may be the smallest part of smart devices like smartphones, smartwatches, tablets, etc. Moreover, it can be minute sensors implanted in a chair, microwave, pen, spectacles and almost any item of daily use [2]. These things are mostly resource-constrained and can perform tasks such as sensing, monitoring, pre-processing, and exchanging of data.

The first real concept of IoT was invented in 1982 by the Students at Carnegie Mellon University. It was truly the first practical application of IoT, where a Coca-Cola vending machine at university was connected to the main computer with Advanced Research Projects Agency Network (APRANET) to ensure if the loaded drinks are cold or not. Later, in 1990, an IoT device-based toaster was invented by John Romkey, which again got automated in 1991. In 1999, Kevin Ashton presented an idea of IoT where he innovatively linked RFID tags with the internet and in the 2000 and then an internet-enabled refrigerator launched by LG was a big sensation in the IoT global market. Thereafter, till date we are witnessing huge innovation and growth in IoT-driven networks [3].

Now, let's discuss about other important aspect of BCoT, that is, blockchain. If we look at it from a layman's perspective, we can say that Blockchain is a rapidly spreading, locked and shared record keeping ledger in which a copy of this ledger is kept at each user node and it can be changed only if all the users agree to do that.

On the other side, the technical perspective says: A peer-to-peer, distributed, immutable, append-only and decentralized ledger which is secured with cryptographic techniques and updateable only via agreement or consensus among parties or peers [4].

Blockchain technology has now progressed into a crucial management solution for all variety of global industries [5] whereas originally it was only related with the first cryptocurrency Bitcoin, founded in 2009 and non-fungible tokens (NFTs). The early projects like CryptoPunks and Rare Pepe were examples of this, though they were not yet widely referred to as NFTs. Peer-to-peer distributed property is a significant reason to combine this technology with IoT which directs us toward BCoT which is the topic of discussion in this chapter. Secondly, cryptographically secure and immutability is the answer to the trust and security concerns in IoT networks.

IoT often lacks the security and authentication which is needed to keep user information secure and private. Presently, we are witnessing betterment in standards of living. It has been possible due to digitization of everyday life with the help of IoT. Technically, we can say that it is the magic of using big data extensively as well as diverse use of smart appliances and sensors [6]. After exploiting IoT functionality, concern raised is a user's safety. There is an urge for high security, ease of maintenance along with standards for IoT and connected devices. All of this can be achieved with introducing Blockchain to the IoT network. Distributed ledger in Blockchain facilitates the data tracing captured by sensors and actuators. Additionally, it blocks the cyber attackers from duplicating that data with other malicious forms of data. Traditionally, IoT uses third parties to exchange information and data, the third-party system opens up even more doors for data and information to be breached, but use of blockchain removes this necessity. Also, costs incurred by third parties are avoided by BT.

Moreover, businesses in the market today are utilizing the concept of Internet of Things (IoT) to explore and modify various ways to exploit the functionality of sensors, intelligent edge devices, and smart infrastructure. As we know the IoT ecosystem functions as a leveled structure, information protection is a major challenge at all levels. With the exponential growth in the number of IoT devices every day, data security is another burning issue. There is an urgent need to resolve this and blockchain serves this.

11.2 ROLE OF BLOCKCHAIN IN IOT

According to recent reports by precedence research, in 2021 the blockchain in IoT market size was valued at $134.41 million and is expected to reach $1.974 billion in 2030, experiencing a CAGR of 73.5 percent from 2022 to 2030 [7]. The Blockchain–IoT development market is experiencing swift growth due to the increasing acceptance of IoT technologies. Businesses and industries acknowledge IoT's potential to improve operational efficacy and enhance security. As a result, there is a growing demand for BCoT to address the challenges imposed by IoT devices.

11.2.1 Data management and data provenance

BC can manage data generated in the IoT sector effectively due to its decentralized and immutability nature. The Blockchain is a distributed electronic database of digitally signed transactions that are clustered into chain blocks [8].

In today's fast-growing era, automated exchange of data between devices without human intervention, which is usually referred to as machine-to-machine transactions is the need of the hour and is served by BCoT. This is made possible with a distributed database with a set of transactions usually referred to as mempool in BC technology, validated by multiple nodes and recorded in a common distributed ledger, which is shared across every node in the Blockchain network. BC–IoT allows machines to communicate and make transactions without relying on a central authority. This means that smart devices can operate independently, making their own decisions. Data provenance can be enhanced with decentralized identity (DID) provided by BC, which maintains info about which device generated the data and who accessed or modified it. Another property of BC is a smart contract, which is effective in tracking the data flow in IoT networks. This smart tracking also strengthens provenance. For example, a smart contract can automatically log the transference of data from a device to a cloud server, recording every step in the data lifecycle.

11.2.2 Enhanced security

IoT enables multiple devices to access, exchange, and share data within the ecosystem. However, without human administration and authentication guidelines, an IoT device is vulnerable to hackers, it can serve as an entrance point into the network and cause private and sensitive data at stake. Some types of security concerns possibly exist are software attacks, physical devices attack, network attacks, and encryption attacks. As IoT devices increasingly handle sensitive data, the role of Blockchain in bolstering IoT security is becoming more prominent. In this situation, past usage of all the linked IoT devices and sensor data is kept in-check by BCoT, which helps with maintaining the track in real-time. Here the word 'track' is intended to monitor for content deceit from threatful data sources, and also to guarantee security benefits along with identity validation. Transactions are validated and recorded across multiple nodes, making it difficult for any malicious activity or tampering to occur due to the decentralized nature of the blockchain. The immutable nature of Blockchain technology provides an additional layer of protection against malicious activity and can easily safeguard the Blockchain in IoT security and confidentiality. By utilizing advanced cryptographic algorithms in BCT guarantees integrity and authenticity of data transmitted between IoT devices and prevent unauthorized access. Industries like healthcare, supply chain management, and energy, where the exchange of sensitive information and real-time communication are critical components, can benefit from using resources that employ BCoT.

11.2.3 Liquidity provision

Blockchain is capable of offering the means to create digital tokens or cryptocurrencies that can represent asset ownership. These tokens can be utilized within the IoT environment by allowing IoT devices to participate in tokenized ecosystems. BC liquidity supports efficient micropayments, decentralized data trading, and making it easier for businesses or individuals to trade this data in decentralized marketplaces. Liquid markets for tokens ensure continuous trading and pricing transparency. To understand applicability of liquidity in BCoT, we can think of examples such as a smart car could pay for tolls or charging stations in real-time using liquid digital assets or a farmer could sell IoT sensor data to weather prediction agencies using liquid tokens or stablecoins. Reward systems or crowdsourced IoT systems can be implemented with utilization of BC liquidity. BCoT allows for a more rationalized and effective process of value exchange, thereby enhancing the overall functionality and effectiveness of the IoT ecosystem.

11.2.4 Trustworthy/transparent environment

Lack of trustability and transparency is the prevailing challenge in the IoT marketplace. There is usually a higher risk of collision and distrust, as every digital device can connect to the IoT network and gain access to information independently. BC can fill this gap and add trust to the IoT ecosystem by leveraging the characteristics of transparency, immutability, and decentralization. BC employed the public ledger of transactions can be utilized by IoT that is tamper-proof and cannot be manipulated. This achieves transparency among record and storage of transactions along with data exchanges in an IoT network. By keeping track of how devices interact with each other, BC provides a transparent environment, which allows the BCoT devices to work together seamlessly and efficiently. Furthermore, the fusion of these two technologies also helps increase data accountability by tracking the origin and movement of goods across the IoT network.

11.2.5 Improved scalability

Since IoT systems operate via centralized architecture, connecting billions of devices into the same network is becoming increasingly tougher, an impact of which is the rise in network issues and cyberattacks. BC enables the integration of multiple IoT systems, irrespective of their hardware or software configurations. This aids in improvement in scalability and development of complex IoT applications.

Layer 2 scaling functions of Blockchain like state channels and roll ups also play a vital role in scalability of BCoT network. State channels enable off-chaining, which allow users to conduct multiple transactions outside the Blockchain and only settle the final result on-chain, as a single batch, reducing network congestion and transaction costs, ultimately helping in improving scalability.

Secondly, sharding concepts in BC helps to employ parallelism. Separate shards managing various device groups can reduce latency and improve performance rising scalability in BCoT.

11.2.6 Low processing overhead and cost savings

In conventional IoT systems, intermediaries or central authorities are frequently required to oversee transactions and validate data authenticity. There are numerous third parties involved that are responsible to guarantee security in the IoT network by checking and authenticating each and every micro-transaction. This is not only a prolonged process but also an extravagant expense. By utilizing Blockchain technology, to eliminate the need for intermediaries, smart contracts can be efficiently employed. This results in reduced costs and minimized delays in transaction settlements by introducing the power of automation. Also, BCoT makes it possible for the removal of unnecessary layers of communication, simplifying the complex IoT ecosystem.

11.2.7 Enhanced privacy

Blockchain technology gives users increased authority over their data, enhancing overall privacy. BC in IoT provides privacy by providing decentralized data management, access control, and introducing anonymity. Access control allows users to determine which information is shared, who has access to it, and the specific conditions under which it can be accessed. It allows individuals to safeguard their data and comply with relevant data protection regulations. Distributed ledger provides authorized access promoting privacy. Moreover, efficient usage of cryptographic techniques such as zero-knowledge proofs (ZKPs) by BC also creates a level of privacy in the IoT environment.

11.3 CURRENT TRENDS IN BCOT DEVELOPMENT

This section will focus on current trends such as increasing use of AI and ML with BC–IoT, privacy-preserving techniques such as zero-knowledge proofs (ZKPs), and lightweight blockchain protocols such as DAG. A slew of prior research is being considered that examines the various current trends in BCoT. There is no denying the many advantages that blockchain offers in the IoT domain. We have discussed a few current trends in this section. These trends are transforming applications across industries, offering solutions to some of the core challenges in IoT environments which we have already discussed in prior sections.

11.3.1 Zero-knowledge proofs protocol (ZKPs)

Zero-knowledge in cryptography first appeared in the 1985 paper "The knowledge complexity of interactive proof systems [GMR85]" by pioneers Shafi Goldwasser, Silvio Micali, and Charles Rackoff. ZKP cryptographic protocols served two parties to communicate and authenticate each other. These two parties are referred to as Prover and Verifier. Prover can prove a certain thing to the verifier without sharing additional details related to it and verifier can validate it [9]. In simpler terms: "I know X, which I'm not going to tell you, but I can prove to you that this statement involving X is true.".

ZKP holds three standard properties:

- *Completeness*: If the prover succeeds to prove the statement true, the verifier has to accept it.
- *Soundness*: If the statement is not true, systems work in such a truthful manner that a dishonest prover cannot prove it true.
- *Zero-knowledge*: There is zero leakage of information from the interaction of both the parties. Only learning happens is the statement given by the prover is true.

Zero-knowledge proofs-based techniques for verifiable identity in the internet of Things were proposed by an author Wohnert et al. [10], highlighting integrity and accountability. With amalgamation of ZKPs with Groth16 proof generation mechanism and Ethereum smart contracts, access control mechanisms in IoT can be enhanced [11]. Groth16 is a zkSNARK protocol introduced by Jens Groth in 2016 and it saw an early application in ZCash. Its smallest and fastest ZKP. Additionally, while focusing on accountability in Blockchain identity management, managing privacy preservation is also of utmost importance, as claimed by research recommended for the traceability factor in ZKP-enabled systems [12,13]. By using ZKPs in the network, IoT devices can exhibit the specific link among functions and data, which is essential to guarantee the integrity of data as well as function. Here the function implies the firmware used in the IoT network, which is responsible for the device's core functionality, including how it interacts with sensors, how it communicates, and how it processes data.

Furthermore, ZKP can be extended to anonymous communications, anonymous payments, and anonymous access. ZKP-based schemes can also avoid forgery, password attacks, replay attacks, tracking, etc. by provision of *Man-in-the-middle Resistance* [9].

11.3.2 Lightweight Blockchain protocols

Lightweight Blockchain protocols particularly focus on reducing resource consumption, improving scalability, and ensuring energy efficiency, which makes them suitable for IoT environments where devices often have limited computational power, storage, and battery life.

11.3.3 IoTA

Internet of Things Application is a decentralized protocol specifically designed for IoT, using a directed acyclic graph (DAG) structure called "Tangle" instead of a traditional Blockchain.

This protocol does not need validation externally. Transactions are validated by the sender themselves by confirming two previous transactions. No need for any consensus mechanism, which makes it scalable and efficiently suitable for IoT. Smart cities, supply chain tracking, and IoT-based microtransactions are some of the case studies which employ this protocol.

11.3.4 Hierarchical Proof-of-Capability (HPoC)

HPoC is a newer Blockchain consensus mechanism tailored for IoT. It is designed to reduce latency and energy consumption, addressing the constraints of IoT devices. It is a lighter Proof of Capability consensus, which can federate the computing, storage, and network bandwidth resources of IoT–edge devices. It can be combined with asynchronous Proof of Work consensus to improvise issues like latency and energy consumption in block generation. It also proved to be cost-effective. Additionally, node storage load can be reduced by using asynchronous Proof-of-Work consensus with a hierarchical pattern. Here, hierarchical pattern implies that nodes are not equal, they are organized as per their storage capacity, computational strength, and other performance attributes. Also, the consensus can be executed at various intervals; for example, every minute, each ten minutes duration or per day, which will further improve security of Blockchain data against unauthorized access, modification, or interference [14].

11.3.5 LightBlock

LightBlock integrates lightweight and scalable Blockchain techniques optimized for IoT applications, ensuring low computational demand and high throughput while maintaining privacy and security [15]. LightBlock uses mainly three components: SBA, LCA, and TMS. SBA is a scalable blockchain architecture, LCA is a lightweight consensus algorithm, and TMS is a throughput management scheme. Lightblock uses functionality of POS (Proof of Stake) consensus, Ethereum 2.0, and DApp (decentralized applications). The first element helps to reduce network overhead, whereas the second element Ethereum 2.0 improves sustainability in the network. The ability of DApps to handle multiple transactions at once makes it most suitable for applications handling voluminous transactions. This property is highly effective for gaining high throughput. These amalgamations make the network highly secure against all the attacks which is utmost important in an

IoT application. The Lightblock can be implemented on various networking topologies using REMIX IDE and Ganache platform.

11.3.6 zk-Rollups

Ethereum's recent push toward scalability with zk-Rollups provides an efficient Layer-2 solution for IoT systems by enabling off-chain data processing and on-chain security verification. It also handles computation like hashing as well as stores the state off chain, which ultimately improves the throughput on Ethereum Mainnet [16]. ZK-rollups is master in handling millions of transactions in a single batch and then summarizes the data to upload on Mainnet. Summarizing the data can lighten the network load. Users benefit from greater capital efficiency and can withdraw funds from L2 without delays. Users don't have to validate the chain to protect their funds. Here, the additional cost incurred for call data published on Ethereum can be handled by employing various data compression techniques. This can also reduce the rollup fees paid by users. Additionally, compressed data can lighten the communication overhead on the network channels.

11.3.7 Use of AI and ML with BC–IoT

Agriculture, healthcare, and autonomous vehicles can be some of the case studies, which can give an idea about integrating AI and ML features with BCoT [17].

K-Means is applied to cluster IoT devices based on similar characteristics or behaviors, such as geographical location or usage patterns. Blockchain stores clustering results securely, ensuring they are auditable and cannot be altered. Smart cities application employs this.

Support vector machines (SVMs) classify IoT data into categories, such as detecting normal versus anomalous patterns. Blockchain acts as a log of classification results, ensuring trust in the decisions made. Example scenario is security systems for IoT devices to detect anomaly behaviors.

Natural language processing (NLP) algorithms process unstructured IoT data, such as voice commands or textual inputs. BC Ensures secure, auditable records of NLP outputs and decisions. The application scenario is smart homes.

Deep learning algorithms such as convolutional neural networks (CNNs) and recurrent neural networks (RNNs) process large IoT data streams, identifying patterns such as anomalies or fraud. Blockchain ensures secure data provenance for the processed IoT data.

For instance, in supply chain management, DL models detect anomalies in sensor data, while Blockchain ensures tamper-proof recording of detected events.

11.3.8 BaaS

IBM debuted the ADEPT system in 2015 as a distributed, Blockchain powered Internet of Things proof of concept. This is the classic example of early use of Blockchain with IoT. This system was built on Bluemix and developed in association with Samsung. Blockchain as a Service (BaaS) is a cloud-based system practiced by companies to implement Blockchain-enabled IoT apps. Initially in 2021, Microsoft was offering a Quorum Blockchain-as-a-service on the Azure Cloud, which is now handled by ConsenSys, a third party, to customers of that service. Trending startups culture is a pioneer to explore this product. Amazon Web Services (AWS) is also ruling that Blockchain as a profound solution for those exploring to get on the Blockchain train. By simply pressing a few buttons in the open Ethereum Blockchain network, users can swiftly start a node. Fog and cloud are the most convenient platforms to host BaaS. These platforms can permit users to implement and deploy services online.

Smart contracts, smart apps, or other digital services enables secure, transparent and automated solutions in various domains such as supply chain traceability, digital voting, retailer loyalty programs, and real estate transactions reducing the need of intermediators. To address complexity in deployment and maintenance of Blockchain applications, Blockchain-as-a-Service (BaaS) serves as a perfect solution. This also eases developers to focus more on implementation of business logic. This has boosted the practice of BaaS to simplify implementation, deployment, and maintenance of BC applications to the next level. Microsoft, Amazon, R3, and SAP Cloud Platform blockchain are the major tech firms which are active vendors for BaaS in market [18].

11.3.9 Platforms used for BCoT

Today, the Blockchain-enabled IoT market mainly demands the benefits of high security, high privacy, and functional efficiency as it adapts to the inclusion of the IoT in various applications. Following are some of the platforms utilized to implement Blockchain technology-assisted IoT applications.

11.3.9.1 IoTA

IoTA was built in 2015 as a very first distributed ledger for IoE, that is, Internet of Everything. This exploits the functionality of tangle, the directed acyclic graph to enable processing of several transactions in parallel, which is not the case in traditional BC. IoTA serves as an integration stack, which is a base for a digital economy. With less computing power along with limited resources, high throughput in terms of reliability and high number of block validation is made possible with this trending tech [19]. IoTA wallets, identity, tokenization, and tooling are some of the advanced services built upon IoTA.

11.3.9.2 IoTIFY

It is observed that developing IoT-based solutions takes much time. IoTIFY is a solution for this problem. It is a cloud-enabled simulator, which is scalable, flexible, and easy to use. It can handle millions of IoT devices smoothly and aids in quick development, testing and deployment with minimum hassle [20].

11.3.9.3 iExec

iExec is a Blockchain-based tool, which is available on an open platform. It serves as an ecosystem for privacy-preserving applications which are autonomous and decentralized in nature. Basically, it aids virtual cloud infrastructure by using Ethereum smart contracts which is beneficial to provide on-demand services with high-performance computing [21].

11.3.9.4 Xage

It is an access control mechanism with zero trust [22]. This provides resilience and robustness against several cyberattacks.

11.3.9.5 SONM

Supercomputer Organized by Network Mining is a decentralized fog computing platform that uses Blockchain to provide cost-efficient computational power. The main objective of SONM is to generate a free marketplace where distributed, decentralized, and hardware-based cloud-based services can be supplied.

In today's digitized world, everyone wants to connect everything so that effective communication can happen in real time. The other expectations in such a scenario are privacy, security, integrity, and authenticity. The amalgamation of BC and IoT satisfies all of them [23]. This integration achieves human-less communication. SONM framework practices efficient and cost-effective communication where data gets transferred at the correct location, in the correct format and in the correct manner, that is, in a secure way. SONM is much similar to projects Golem and iExec as all three are focused on serving a globally distributed marketplace of computing power. In SONM, use of fog computing over clouds furnishes a much safer and cost-effective outcome [24]. Video and CGI rendering, web hosting, scientific applications, online game servers, and neural networks are some of the well-known use cases of SONM.

11.4 CHALLENGES AND RISKS

Going by what we have discussed so far, the potential advantage of BCoT is the autonomous functioning of smart devices. Adding distributed ledger also removes the necessity of the centralized authority. Communication among devices can be easily traced, but sometimes due to the decentralized nature of Blockchain, IoT can face a problem. IoT platforms usually employ client–server architecture or hub-and-spoke model, where spokes are peripheral devices and hub is a centralized entity. Forming a decentralized IoT platform will be compatible with BC technology, but configuring IoT sensors and actuators to handle their own computer and data storage, for computing and data handling will be challenging as IoT devices themselves hold neither computing capability nor storage. High processing power required for Blockchain transactions can impact the speed of resource-constrained IoT devices.

The adoption of IoT–Blockchain technology is not yet widespread due to technical concerns and operational challenges. Steep learning curves associated with different technical aspects of Blockchain can be a major reason for this. Scalability issues due to Blockchain can be faced in IoT network, as BC computation requires huge storage. Adding Blockchain makes IoT secure but it can also introduce new vulnerabilities like Sybil attack and 51 percent attack, which is common for Blockchain. Another point of discussion can be heterogeneity and diversity in the IoT network. As we know, several heterogeneous physical devices are interconnected with the IoT network, which can cause inconvenience in integration with BC. It may need additional middleware, which further leads to additional cost incurred. Regulations and compliances related to Blockchain also add additional complexity to this amalgamation.

Before employing BCoT in any enterprise, it is necessary to create awareness among customers of prerequisites and prospective snags. Careful assessment before investment is utmost important. Block mining associated with BC consumes high power and electricity, so sometimes it can result in unrealistic power consumption and further operational cost. Some of the use cases with most suitability of BCoT are supply chain management, retail industry, and smart cities. With this discussion, we can say that power consumption, performance, and security can be considered to set appropriate trade-off to federate BCoT in any industry.

11.5 FUTURE TRENDS

This section will include discussion on tokenized IoT ecosystems, prominent integration of advanced AI with Blockchain in IoT.

Globally, the Blockchain-enabled IoT industry is steered by the augmented endorsement of the utilization of IoT, emerging importance on viable efficiency, privacy preservation, and also demanding need for the security in IoT networks.

The growth in this industry will also be pushed by the following:

- Technologizing of business norms and processes
- Use of Blockchain technology to serve high transparency
- Employing distributed ledger technologies to use smart contracts
- Rapid adoption of Blockchain technology in the healthcare industry
- Amalgamation of trending technologies with smart investment by government agencies.

There are multinational companies who aim at research and development to rule the global BCoT market. Some of the leading companies claiming in global Blockchain–IoT market are listed here:

- Microsoft
- IoTEX
- Amazon
- Ethereum Foundation
- Intel
- Cisco Systems
- IBM
- KrypC
- The Linux Foundation

BCoT is getting adopted by large enterprises as well as small and medium enterprises (SMEs). Market segments covered by this adoption along with application -wise and end industry-wise list are given below:

BCoT practices in key market sectors:
- Component-wise
- As per hardware configuration
- As per services
- As per software

Based on application
- Requirement of data security
- Use of smart contracts
- Asset tracking and management
- Need of data sharing

Based on end-use Industry
- Construction
- Transportation (automated vehicles)

- Consumer electronics
- Smart city applications
- Healthcare sector
- Manufacturing
- Retail industry

11.5.1 Tokenized IoT ecosystems

Tokenized IoT ecosystems in BCoT can involve the use of digital tokens to manage and incentivize IoT operations, enhancing automation and transparency. Data monetization can be employed by tokenizing valuable data generated by IoT devices and using Blockchain, it can be traded securely over the network. Autonomous vehicles can pay tolls by utilizing microtransactions by realizing tokens facilitated real-time payments between IoT devices. Tokens can help to decentralize energy markets; for instance, tokens used for energy trading among smart grids.

11.5.2 Integration of advanced AI with Blockchain in IoT

- Transfer Learning with BCoT

The extensive utilization of actuators and sensors throughout the whole manufacturing area has markedly altered how humans indulge with some of these cumbersome processes transfigure them into intelligent environments that can acknowledge their existing status and take relevant action. Transfer learning is a technique where learned knowledge can be applied to do another task in a related context. When we talk about the machine learning domain, usually problems to be solved are limited to a single domain, that is, training and testing data used to train models are with the same distribution. However, practically if the distribution changes even with minor quantities, the model needs to be retrained. This is cost oriented and infeasible to achieve. In such a scenario, TL comes handy. The main application of transfer learning can be in IDS, that is, intrusion detection system in IoT domain. It can be used to detect components in industrial machinery or anomaly detection in sensor data.

Transfer learning also has a good application prospect in the industrial IoT environment. Many training tasks in the IIoT have certain interoperability. The Industrial Internet of Things (IIoT) is a paradigm shift, primarily in the domain of manufacturing industry [25]. If transfer learning is applied to these tasks, it will save a lot of time and computing resources. So BCoT and use of transfer learning technology can improve the overall performance of the IIoT system.

- Federated Blockchain in IoT

One of the recent and profound innovations in Blockchain is federated Blockchain. Instead of a decentralized network of nodes, a set of predecided entities govern the Blockchain. These governing nodes have authority to make all the decisions about overall governance of Blockchain. Low transaction cost, faster speed, scalability, low energy consumption, and no risk of criminal activity and defense against 51 percent attack are the major positive aspects of federated BC. In short, these are speedier (higher scalability) and provide greater transaction privacy. Federated Blockchain examples include R3 (banks), EWF (energy), B3i (insurance), Corda, etc.

Federated BC has lightweight consensus as it uses a consortium of preselected nodes, which makes it ideal for IoT devices with limited resources. By limiting consensus only to trusted IoT nodes, it can minimize delays and improve throughput which is highly applicable for high-volume IoT networks [26].

- Ricardian Contracts Usability in BCoT

If we want to add a legal aspect to the smart contract, the Ricardian contract is the answer. It is a smart contract with legal agreement and legal framework. A program can be written to lay out all the legal facts as per the Ricardian contract and executed. These contracts are cryptographically signed and verified. The concept of the Ricardian contract was found by a school student named Ian Grigg. Beauty lies in the fact that it is readable to humans as well as computers, so it can exploit benefits of the judicial system's protection and modernism of BC technology. Additionally, objectives and related behavior is registered before the agreement is executed. A well-known application of this type of contract is OpenBazaar online marketplace, where asset transfers are done without scams and with legal protection.

We can say that it is a type of smart contract that combines legal agreements with code. It provides human–machine duality, which means those are understandable by both humans for legal compliance and computers for automated execution. This feature can be useful to formalize agreements between devices and services, ensuring verifiable compliance with policies while automating operations through Blockchain. IoT supply chains, HealthcareIoT, and smart cities can be some targeted domains for this.

- Edge AI in BCoT

As BCoT is growing fast, all developing industries will be demanding speedy decision-making, augmented automation resulting in increased efficiency across all domains. The merging of AI and edge provides results faster, more secure data processing, and removes dependency on cloud-based infrastructure. This causes reduction in network congestion and allows devices to function offline with limited internet connectivity.

Edge AI strengthens industries by providing real-time analytics, enhanced safety, and refining operational efficiency. It has proven its potential to revolutionize multiple fields by showcasing its application in autonomous driving, smart cities, agriculture, and industrial automation. The application of BCoT in the context of Edge-AI not only upgrades the performance but also resolves the issues related to data privacy and safety, modeling it a prime component in the era of AI-enabled technologies.

11.6 APPLICATIONS OF BCOT

Now that we have studied what blockchain means to IoT, let's look at some of the Blockchain–IoT applications across industries.

Some of the well-known industries that are utilizing amalgamation of Blockchain and IoT are banking industry, financial services, automobile industry, and agriculture. Moreover, smart homes, logistics, and supply chains are also the newly exploring areas where BCoT is performing very well.

11.6.1 Supply chain

As we know, the complexity level of the global supply chain process due to which it becomes extremely time-consuming and with less-to-no transparency. The main reason behind this is it involves many stakeholders like suppliers, distributors, customers, various governmental and regulatory bodies, brokers, retailers, and raw material providers. It also involves many payments, transactions, and invoices. It complicates the end-to-end visibility. Besides, the risk of delivery delays is also common. So, to introduce transparency with ease of payments, adopting Blockchain in the supply chain management can be a promising solution. The decentralization nature of Blockchain removes the need of intermediaries which ultimately saves time and cost, resulting in the supply chain as a true ecosystem. Blockchain also serves secure data transfer and tamper-proof repository which is crucial for cargo documents and shipping events in the supply chain. This empowers businesses in a real manner, resulting in fewer delays and fraud. As delivery delays have become the biggest challenge due to the involvement of several stakeholders as discussed before, many companies are practicing utilization of IoT-enabled vehicles to track the movement of products throughout the shipment process. The amalgamation of Blockchain and IoT can address the critical challenges of lack of transparency and complexity in the current supply chain and logistics which can further strengthen the network's traceability and reliability.

11.6.2 Smart homes

Smart homes are one of the finest Blockchain–IoT use cases that have redefined the fusion of two technologies. Various smart home automation companies in the global market like Samsung SmartThings, Apple HomeKit, Amazon Alexa, Google Home, and Philips Hue are embracing blockchain technology to introduce immutability in the data assembled and distributed by smart devices and prevent the vulnerability to hacking.

One of the well-known Blockchain–IoT utilizations is Comcast. Comcast is exploring and implementing Blockchain technology within its IoT solutions, particularly in the connected home space, to enhance security and privacy. They are using Blockchain to enable secure and controlled access to IoT devices, allowing users to grant and revoke permissions remotely. A permission-based ledger is used for this purpose. The beauty of this solution is there's no time barrier for these permissions; it can be invalidated at any time.

11.6.3 Medical industry and health economy (healthcare)

Healthcare and pharmacy are other industries that are revolutionizing with the adoption of Blockchain and IoT. The traditional health industry faces daunting challenges like deficit of data security, prolonged wait, data inconsistency, and escalating prices.

The two technologies are helping the industry mitigate its traditional challenges like lack of data security, longer wait time, data inconsistency, and higher attached costs.

The decentralized, distributed and immutable nature of Blockchain technology relieves the severe hit of clinical research and pharmaceutical development and insecure patient medical verification whereas the introduction of IoT in healthcare guarantees end-to-end connection and data exchange in real time.

Moreover, blockchain employs smart contracts by replacing third-party intermediaries, which fastens up the overall process. IoT usage in healthcare allows the medical professionals and pharmacists to get conversant with any undetermined medical symptoms and prepare for new chronic diseases swiftly and in a well-organized manner.

11.6.4 Automobile manufacturing

The auto industry is also one of the vital applications of BCoT. The two disrupting technologies helped to coin the term "smart parking" by smoothing the act of finding a free space for parking along with use of a crypto wallet which automates the parking payments, eliminating the need for manual payment processes and reducing the risk of errors or fraud. Another one of the famous Blockchain–IoT applications includes RFID tags. RFID is radio frequency identification technology that uses radio waves to trace the exact location of any particular car in the IoT-connected world. Automobile manufacturers are using distributed ledger to ensure proof of provenance for the spare auto parts to check the origin and authenticity. Here, proof of provenance implies to track a spare auto part's journey from its source to the current owner, providing transparency and trust. With publicly distributed ledger, Blockchain can store and share this information globally. The power of smart contracts employed in Blockchain of Things (BCoT) enables customers to claim and process vehicle insurance promptly and in a much easier way.

11.6.5 Technology-driven finance (fintech)

Again, fintech is also experiencing a significant change in its working environment with the advent of BCoT. Here, fintech can be referred to describe technologies, services, and companies in the financial sector that focus on various capabilities, including retail banking, financial education, fundraising, cryptocurrencies, investment management, and more.

If we consider the role of Blockchain in fintech, Blockchain simplifies things by automating the operational processes. The main benefits can be discussed as below:

- Increased process efficiency
  - Reduced friction in cross-system transactions
  - Fewer intermediaries for streamlined, cost-efficient operations
  - Decentralized structure facilitates partner interoperability
- Robust data management
  - Immutable ledger like features ensure data integrity
  - Renowned capability to attest data authenticity and provenance
  - unique identity verification methods to match actions to their actors
- Enhanced transparency and security
  - Independent data verification builds trust of stakeholders
  - Tamper-proof systems reduce the risk of fraudulent activities
  - Auditable nature facilitates regular compliance

Simultaneously, IoT is serving the fintech industry by providing many advantages such as enhanced customer experience with operational efficiency and personalization services, real time data insights, providing spending visibility and assessing risk related to insurance and loan. IoT-powered payment terminals and digital wallets enable seamless contactless payments, reducing dependency on cash and physical cards. Furthermore, blockchain supports payments between devices, making the web of devices immune to hacking and malware attacks.

11.6.6 Sharing economy

The sharing economy is another blockchain–IoT use case that is gaining huge momentum in integrating both technologies. The technologies enable companies to create a universal sharing network (USN) where users could interact directly with each other to rent or share resources using smart contracts. Here, smart contracts are integrated with IoT devices. For example, a smart lock could be programmed to unlock an apartment only when a payment is received via the Blockchain. In today's market, Slock.it is a company that aims to revolutionize the sharing economy by leveraging BCoT. Slock.it envisioned applications like automated rental apartments, on-demand smart object rentals, and the repurposing of underutilized vehicles. While Slock.it's vision was ambitious and aimed to disrupt the sharing economy, the concept of using Blockchain and smart contracts to enable decentralized sharing remains a relevant area of exploration within the broader sharing economy landscape.

REFERENCES

[1] E. Oztemel and S. Gursev, "Literature review of industry 4.0 and related technologies", *Journal of Intelligent Manufacturing*, vol. 31, no. 1, pp. 127–182, January 2020.

[2] F. Restuccia, S. D'Oro, and T. Melodia, "Securing the Internet of Things in the age of machine learning and software-defined networking", *IEEE Internet Things Journal,* vol. 5, no. 6, pp. 4829–4842, June 2018.

[3] S. Mathur, A. Kalla, G. Gür, M. K. Bohra, and M. Liyanage, "A survey on role of blockchain for IoT: Applications and technical aspects", *Computer Networks*, vol. 229, pp. 109726–109772, 2023.

[4] *I. Bashir, "Mastering Blockchain", Second Edition, Packt Publishing.*

[5] G. Habib, S. Sharma, S. Ibrahim et. al. "Blockchain technology: Benefits, challenges, applications, and integration of blockchain technology with cloud computing", *Open Access, MDPI*, vol. 14, no. 11, pp. 341–362, 2022.

[6] A. Bikos, and S. Kumar, "Securing digital ledger technologies-enabled IoT devices: Taxonomy, challenges, and solutions", *IEEE Access, Multidisciplinary Open Access Journal*, vol. 10, pp. 46238–46254, 2022.

[7] S. Zoting, "Blockchain IoT Market Size, Share and Trends 2026 to 2035", www.precedenceresearch.com/blockchain-iot-market, 2025.

[8] D. Mingxiao, M. Xiaofeng, Z. Zhe, W. Xiangwei, and C. Qijun, A review on consensus algorithm of blockchain. In *Proceedings of the 2017 IEEE International Conference on Systems, Man, and Cybernetics (SMC)*, Banff, AB, Canada, 5–8 October 2017.
[9] Z. Chen and Y. Jiang et.al. "A survey on zero-knowledge authentication for Internet of Things", *MDPI, Open Access Review*, vol. 12, no. 5, pp. 1145–1164, 2022.
[10] K. Wohnert, S. Wohnert, T. Thiel, R. Weißbach, and V. Skwarek, "Secure cyber- physical object identification in industrial IoT-systems", *Procedia Manufacturing*, vol. 51, pp. 1221–1228, 2020.
[11] L. Song, X. Ju, Z. Zhu, and M. Li, "An access control model for the Internet of Things based on zero-knowledge token and blockchain", *EURASIP Journal on Wireless Communications and Networking,* vol. 2021, no. 1, pp. 1–20, 2021.
[12] I. Damgard, C. Ganesh, H. Khoshakhlagh, C. Orlandi, and ˚ L. Siniscalchi, "Balancing privacy and accountability in blockchain identity management", in *Topics in Cryptology CT RSA 2021* (K. G. Paterson, ed.), (Cham), pp. 552–576, Springer International Publishing, 2021.
[13] D. A. Luong and J. H. Park, "Privacy-preserving identity management system on blockchain using zk-snark", *IEEE Access*, vol. 11, pp. 1840–1853, 2023.
[14] Z. Nie, M. Zhang, and Y. Lu, "HPoC: A lightweight blockchain consensus design for the IoT", *MDPI, Open Access article*, vol. 12, no. 24, pp. 12866–12887, 2022.
[15] M. A. Mahmoud, M. Gurunathan, R. Ramli, et. al. "Review and development of a scalable lightweight blockchain integrated model (LightBlock) for IoT applications", *MDPI, Open Access Article*, vol. 12, no. 4, pp. 1025–1042, 2023.
[16] https://ethereum.org/en/developers/docs/scaling/zk-rollups/
[17] S. S. Gill, S. Tuli, M. Xu, I. Singh, K. V. Singh, D. Lindsay, S. Tuli, D. Smirnova, M. Singh, U. Jain, et al. "Transformative effects of IoT, Blockchain and Artificial Intelligence on cloud computing: Evolution, vision, trends and open challenges", *Internet Things*, vol. 8, p. 100118, 2019.
[18] M. M. H. Onik and M. H. Miraz, "Performance analytical comparison of blockchain-as-a-service (BaaS) platforms," in *Proceedings of the International Conference on Emerging Technologies in Computing*, Cham, Switzerland: Springer, 2019, pp. 3–18.
[19] IOTA Technology Showcases, www.iota.org, accessed on May 2025.
[20] IoTIFY Network Simulator, https://docs.iotify.io/, accessed on May 2025.
[21] Build with Programmable Privacy, www.iex.ec, accessed on May 2025.
[22] Universal Zero Trust Access, www.xage.com, accessed on May 2025.
[23] R. Brundo Uriarte and R. DeNicola, "Blockchain-based decentralized cloud/fog solutions: Challenges, opportunities, and standards", *IEEE Communications Standards Magazine*, vol. 2, no. 3, pp 22–28, 2018.
[24] T. Alam, "Blockchain-based Internet of Things: Review, current trends, applications, and future challenges", *MDPI Review*, vol. 12, no. 1, pp. 12–41, 2022.

[25] Z. Mahmood, "*The Internet of Things in the Industrial Sector, Security and Device Connectivity, Smart Environments, and Industry 4.0*", Springer, 2019.
[26] J. Song, P. Zhang, M. Alkubati, Y. Bao, and G. Yu, "Research advances on blockchain-as-a-service: Architectures, applications and challenges," *Digital Communications and Networks*, vol 8, no. 4, pp. 466–475, 2021.

Chapter 12

Future of Internet of Things driven by blockchain

Limitations and research scope

Sonal Bankar and Sulbha Yadav

12.1 INTRODUCTION

IoT connects physical items, such as sensors, electrical devices, and appliances, to collect and exchange data online. The IoT ecosystem transforms everyday objects into smart, networked instruments that improve convenience and efficiency in many aspects of life while also changing the way people interact with technology. Smart cities, healthcare, transportation, and agriculture are examples of such fields. As IoT devices are more widely used and data collected by them is enormous, worries about data privacy and security safeguards have grown. Traditional security solutions are frequently insufficient for IoT due to its distinct resource requirements and limits. As a result, there is significant interest in integrating Blockchain to increase the security and trust of the IoT ecosystem. Blockchain promises to address the inherent weaknesses of the Internet of Things due to its decentralized ledger system. It is renowned for its safe, transparent, and immutable transaction features.

Blockchain, first introduced in 2008, was designed as a distributed ledger system that ensures data integrity and transparency. While it initially served as the foundation for digital cryptocurrencies, its unique features have since led to a wide range of applications beyond its original purpose. The technology relies on various consensus mechanisms such as finality, authorization, proof of work, proof of stake, and other consensus models to maintain trust and agreement across the network. Blockchain networks generally fall into three categories: public, private, and consortium. In the context of the Internet of Things (IoT), Blockchain offers effective solutions to several key challenges. It can help with device identification and authentication, enable secure end-to-end encryption, facilitate device-to-device transactions, and ensure data integrity and protection. Additionally, it supports secure device registration, management, and attestation, while also providing a distributed database framework for governing authorities.

IoT and Blockchain are two rapidly developing revolutionary concepts, and their growing applications in business and daily life have also given rise to challenging deployment problems involving security, privacy, and trust.

DOI: 10.1201/9781003584728-12

IoT architectures can adopt the fundamental ideas of Blockchain to sustain dispersed and secure transactions in order to develop a timely automated system. In the next section, the key components of the IoT are outlined, with emphasis on the advantages and disadvantages of this technology. The main challenges are then named, and Blockchain for IoT applications is proposed. The analysis of Blockchain solutions and IoT interactions, along with their advantages and disadvantages, is presented. Suggestions are given for future research in the IoT and blockchain.

12.2 LITERATURE REVIEW

Blockchain technology is a great option to combine with IoT technologies since it offers the fundamental layer for machine-to-machine transaction, secure data transmission, device identity and verification, and data sharing. Examining the possible uses of Blockchain technology in an Internet of Things context can be profitable for researchers and entrepreneurs. A robust union will produce a secure network that can reasonably guarantee privacy, security, and reliability. This combination can give IoT a solid foundation to overcome present and upcoming challenges and accomplish a large-scale deployment objective with sufficient scalability. By facilitating extensive physical entity and item interconnection for innovative services and activities, the Internet of Things, which is at the forefront of technology, established a new paradigm for the Internet. However, decentralized digital currencies established Blockchain technology as the foundation of the technological revolution to capture and protect information about online transactions. The future and continued worth of IoT will be determined by how well it can safeguard users while preserving their privacy, guaranteeing the security and integrity of data flow, and enabling lifelong tracking. Authentication, integrity, confidentiality, and trust delegation being the main goals.

Ayoade et al. [1] propose a system to address IoT data security challenges. It integrates Blockchain for decentralized and tamper-proof data storage with a trusted execution environment (TEE) for securely processing sensitive data off-chain. The approach uses smart contracts for access control and stores hashed data on-chain, while raw data is managed within TEE. This combination ensures data confidentiality, integrity, and auditability, addressing limitations of centralized IoT data management systems.

The authors of [2] examine the structure and workings of Blockchain technology, emphasizing its fundamental elements, including distributed ledgers, blocks, and cryptographic hashing. It looks at consensus techniques like Proof-of-Work (PoW) and Proof-of-Stake (PoS), highlighting how important they are to preserving security and decentralization. The study also explores possible future developments, including acceptance across various industries, IoT integration, and scaling options. While addressing issues like energy consumption, legislation, and interoperability for wider adoption, it emphasizes Blockchain's revolutionary potential.

Alam [3] provides an in-depth exploration of integrating Blockchain with IoT technologies. The study highlights decentralized, transparent, and secure ledger systems in Blockchain as a critical enabler for addressing IoT's challenges, such as data security, trust, and scalability. Blockchain ensures immutable records of IoT transactions and enhances communication between IoT nodes by leveraging its consensus mechanisms and smart contract capabilities. Key aspects include a review of Blockchain-enabled IoT architectures, real-world applications (e.g., supply chain, healthcare, and smart cities), and emerging trends such as edge computing integration. The chapter also identifies challenges, including computational overhead, energy consumption, and interoperability issues, while emphasizing the need for advanced consensus protocols and lightweight Blockchain models tailored for IoT

Makhdoom et al. [4] present a data-sharing architecture that is both secure and privacy-focused, specifically designed for smart city settings. In order to improve data isolation and scalability, it uses a multi-channel Blockchain approach, in which each channel controls distinct data types. Access control is managed using smart contracts, which guarantee that only authorized parties have access to data. To safeguard user privacy and control data usage, the framework integrates encryption, private data collecting, and General Data Protection Regulation (GDPR) compliance standards. Unauthorized access is prevented with an OAuth 2.0-secured REST application programming interface (API). PrivyCoin, a digital token incentive system, promotes safe data exchange between stakeholders. It has been shown that the system addresses issues like trust and transparency in multi-party ecosystems while enhancing scalability and data protection for IoT-based smart cities.

Within hospital smartphone networks (HSNs), the study in [5] presents a system that uses Blockchain technology to improve data sharing and trust management. The model covers important topics including privacy, insider threats, and data security in healthcare IoT systems. It guarantees safe, effective, and transparent data exchange between healthcare equipment and stakeholders by leveraging smart contracts and Blockchain's decentralized, immutable ledger. Additionally, the system supports privacy and trust in real-time patient data sharing by incorporating techniques to identify and neutralize rogue nodes.

Wang et al. [6] proposed to use permissioned Blockchain technology to provide a strong framework for addressing privacy and data security issues in IoT environments. To enable privacy-preserving data sharing while preserving data integrity and access control, the method incorporates cryptographic techniques like secure hash functions and homomorphic encryption.

Banerjee et al. in [7] explore how Blockchain technology might improve IoT security. It lists the difficulties that IoT devices encounter, such as inadequate security measures, constrained processing capacity, and vulnerability to intrusions. To solve these problems, the authors suggest Blockchain as a decentralized, unchangeable, and safe option. The capacity of Blockchain to guarantee data integrity, stop tampering, and offer transparency in IoT networks are important features. They also draw attention to issues with energy efficiency and scalability when combining Blockchain with IoT systems.

Uddin et al. [8] examine cutting-edge Blockchain applications for the IoT, encompassing fields like intelligent transportation, smart cities, and healthcare. Additionally, it looks at how Blockchain can be integrated with Fog and Cloud IoT systems, outlining how it can improve data security, integrity, and operational effectiveness. In addition, the study highlights issues including interoperability, scalability, and energy consumption and talks about possible fixes and future research avenues to get past them. It gives a complete assessment of incorporating Blockchain into IoT environments. Conventional IoT architectures, which depend on centralized cloud services, have a number of problems, including data bottlenecks, privacy issues, single points of failure, and trouble safely updating device firmware. By removing dependency on third parties and lowering risks linked to centralized systems, Blockchain's decentralized and secure nature solves these problems.

Ali et al. [9] offer a thorough analysis of the ways in which Blockchain technology might improve the IoT ecosystem. Privacy, trust, security, monetization, identity and data management are the six main application areas identified by the survey. The reasons for combining Blockchain technology with IoT are also covered, such as facilitating decentralization and addressing drawbacks of conventional centralized IoT architectures. The authors describe Blockchain integration techniques for IoT, examine consensus algorithms and their trade-offs for IoT applications, and contrast them with other decentralization strategies. They examine problems like energy usage, scalability, and security concerns and provide strategies to get over them. The study also identifies unexplored research areas and offers a framework for directing the future incorporation of Blockchain technology into IoT initiatives.

The paper [10] by Issa et al. (2023) reviews the integration of Blockchain with federated learning (FL) to enhance security and privacy in IoT environments. It explores FL's ability to allow decentralized learning on IoT devices while preserving privacy, addressing issues like data centralization and security risks. The paper introduces a threat model and discusses how Blockchain can provide a secure and transparent framework for federated learning, ensuring trust and robustness in IoT applications.

A thorough analysis of the union of Blockchain technology with the IoT, particularly in industrial settings, is provided by Dwivedi et al. [11] It focuses on how blockchain might improve IoT systems' scalability, security, and privacy, especially in industrial IoT (IIoT) networks. The study examines a number of issues, including latency, energy usage, and the requirement for effective consensus algorithms to handle the large amount of data produced by IoT devices. It talks about existing approaches, upcoming developments, and the possible advantages of applying Blockchain to IoT systems to guarantee data integrity, trust management, and decentralized control.

Zyskind and Nathan [12] suggest a system that uses Blockchain technology to improve privacy and secure personal information. The method gives individuals more control over their data by using decentralized structure in Blockchain. The proposed work allows users to control and audit the use of their data by storing data access rights on the Blockchain. Users maintain control over their personal data since the system only permits data sharing when access policies are followed. Additionally, it presents a strong security paradigm that guards against privacy violations even in the case that certain network nodes are corrupted. By using encryption and the Nakamoto consensus technique to make sure that no one party has complete control over the data, this system improves privacy.

The book chapter [13] by Vashisht et al. explores the main privacy and security concerns related to IoT and how Blockchain might be able to help. It talks about how incorporating Blockchain technology into IoT devices might improve user privacy, data security, and safe transactions in dispersed settings. While Blockchain's cryptographic characteristics help to increase confidentiality and authentication across IoT networks, its decentralized structure offers transparency and tamper-proof data storage. Although Blockchain presents promising solutions to these issues, problems like scalability, energy usage, and the difficulties of overseeing a high number of IoT devices still exist. The chapter also highlights the need for more study to create consensus systems and Blockchain designs that are effective in IoT settings.

At the present time, there are billions of smart devices in use that are linked to the Internet and to one another. It is anticipated that this number will increase exponentially in the future. Globally, the number of devices connected to the Internet of Things (IoT) has increased dramatically over time and is predicted to continue growing quickly. By 2023, there were 51.11 billion IoT devices, up from 15.41 billion in 2015. According to projections, the number will increase to approximately 62.12 billion by 2024 and 75.44 billion by 2025. Growing IoT adoption in sectors like manufacturing, healthcare, transportation, and smart homes is what's causing this growth. Additionally, IoT is now dominating the global device landscape, with IoT devices expected to outnumber non-IoT devices by a ratio of 3:1 by 2025. The steady expansion of IoT underscores its transformative impact on connectivity and its potential to revolutionize various sectors globally.

In Figure 12.1 is provided the year-wise statistics from 2015 to 2025 for the global number of IoT connected devices, based on available data.

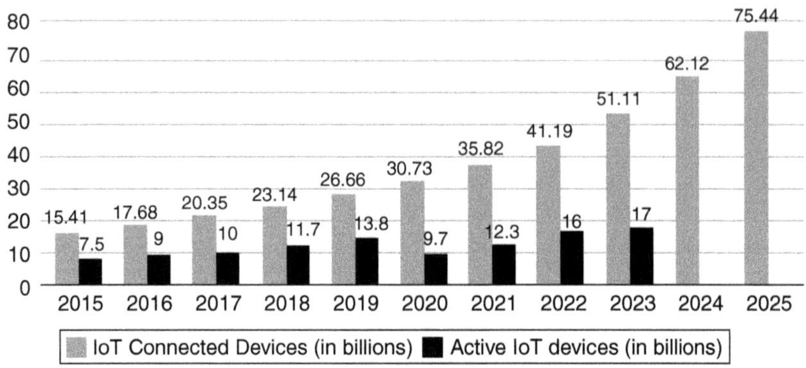

Figure 12.1 Statistics for the global number of IoT connected devices and active devices.

These estimates indicate steady growth in IoT adoption, driven by the need for intelligent solutions across industries, greater connectivity, and technological advancements. According to the statistics, IoT devices are rapidly surpassing non-IoT devices worldwide, indicating exponential growth.

IoT connections are expected to reach 38–40 billion devices, reflecting a compound annual growth rate (CAGR) of about 10 percent.

By 2030, enterprise IoT will account for over 60 percent of connections, with significant growth in smart manufacturing (CAGR of 20 percent) and other sectors such as smart buildings and homes. According to the statistics, IoT devices are rapidly surpassing non-IoT devices worldwide, indicating exponential growth.

Although cellular and low-power wide-area network (LPWAN) [14] technologies, such as 5G, will also significantly grow, short-range connections will remain the most common. Due to its wide-ranging effects, IoT requires a careful analysis of ecosystems, device architecture, and the surrounding environment in order to find any inherent vulnerabilities or complications and figure out how to reduce them.

With the abundance of devices and interconnected networks, the IoT paradigm brings with it traditional security concerns from computer networks. Furthermore, security issues are exacerbated by the limited power and memory of small devices or sensors. Security solutions must therefore be adjusted to function with these constrained architectures.

Table 12.1 provides a summary of the papers. This table summarizes the contributions and focus areas of each paper, highlighting their exploration of the role of Blockchain in enhancing IoT privacy, security and decentralization.

The authors have introduced a framework where Blockchain decentralizes access control, allowing users to manage their personal data securely while

Table 12.1 Literature summary

Paper	*Focus on use of Blockchain*	*Key points*
Zyskind et al. [12]	Personal data privacy	Proposes a Blockchain framework for decentralized control over personal data, ensuring secure access and privacy via cryptographic methods
Issa et al. [10]	Federated learning in IoT	Explores how Blockchain can secure federated learning in IoT, providing transparency, trust, and privacy without exposing data
Dwivedi et al. [11]	IoT and IIoT	Show the role of Blockchain in enhancing security, privacy, scalability, and data integrity in IoT and Industrial IoT
Vashisht et al. [13]	IoT security and privacy	Highlights Blockchain's potential to secure IoT communications, prevent tampering, and ensure privacy through decentralized data storage
Makhdoom et al. [4]	Blockchain for smart city data sharing	Presents PrivySharing, a Blockchain-based framework with an emphasis on user control for safe, private data sharing in smart cities.
Ali et al. [9]	Blockchain in IoT applications	Extensive analysis of Blockchain's application in the IoT for development of decentralized systems, finding management and safe data sharing.

maintaining privacy. The role of Blockchain in securing IoT and Industrial IoT (IIoT), addressing issues like latency and energy consumption, while improving data integrity and scalability are discussed. These papers collectively emphasize Blockchain's potential to address key IoT challenges, although scalability and consensus mechanisms remain critical areas for further research.

12.3 INTERNET OF THINGS

Although widely used, the term "IoT system" is still ambiguous and diverse [15]. The three approaches of thing oriented, Internet-oriented, and semantic-oriented approaches to defining IoT systems are examined by Tran et al. (2021) [16]. Practically speaking, they define IoT systems as tangible objects that connect online via electronic tags, sensors, and actuators. The three-tier architecture used by IoT systems frequently consists of cloud, fog, and edge components.

To increase scalability and performance, the three-tier IoT design separates functionality into edge, fog, and cloud layers. The edge layer is closest to

IoT devices, such as sensors and actuators. In order to minimize latency and bandwidth consumption, it carries out data collecting, preliminary processing, and real-time operations locally. The fog layer is the computing infrastructure that sits in between the edge and the cloud. By processing data closer to the point of origin, it enhances responsiveness and makes tasks like filtering, analytics, and decision-making possible. Cloud Layer offers global insights, sophisticated computations, and centralized storage. It takes care of system-wide coordination, data management, and long-term analytics. In IoT systems, this architecture improves scalability, efficiency, and dependability (Figure 12.2).

The IoT design has a number of problems at all three layers. Starting with the edge layer, major limitations include the limited processing and storage capacity of IoT devices. The second issue is security threats such as device hacking and data interception. The incompatibility of different protocols and devices is one more problem in interoperability. The fog layer faces the challenges of balancing data flow between the edge and the cloud. Organizing dispersed resources across many geographical areas and scalability is a major problem. Security hazards like the possibility of illegal access or data leaking is also there. In the cloud layer, one point of failure could be centralization. The data overload due to the massive volume of data produced by IoT devices and latency due to centralized data handling is

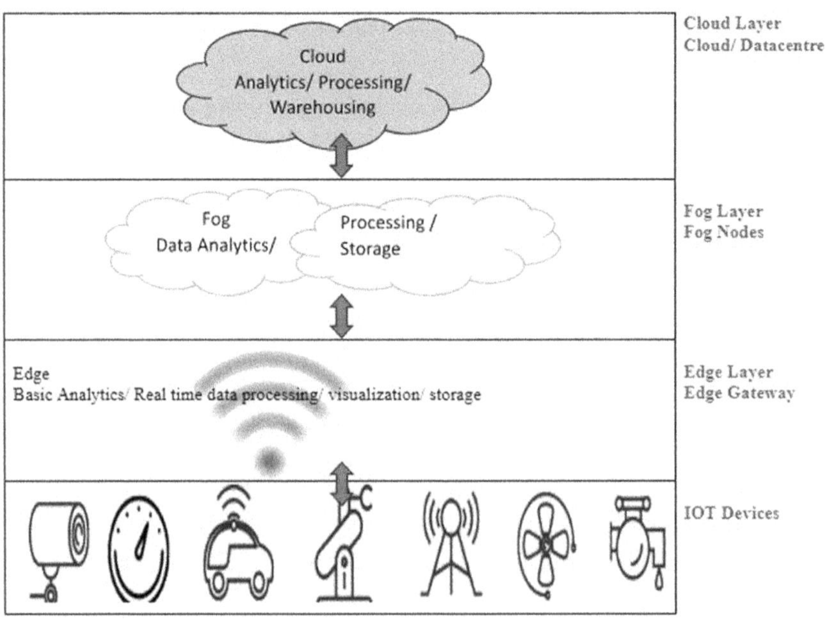

Figure 12.2 Edge–Fog–Cloud architecture.

an issue. To solve these problems, scalability, security, and effective resource management at every layer must be improved.

12.4 BENEFITS OF BLOCKCHAIN AND IOT INTEGRATION

There are numerous revolutionary advantages to integrating Blockchain with IoT (Figure 12.3). They are as follows:

Enhanced security: By using cryptographic techniques to secure IoT networks, Blockchain lowers the susceptibility of these networks to attackers by guaranteeing secure communication and tamper-proof data storage. Blockchain improves IoT system security in a number of ways:

Immutability of data: The immutability of Blockchain records guards against unwanted changes and guarantees the accuracy of IoT data.

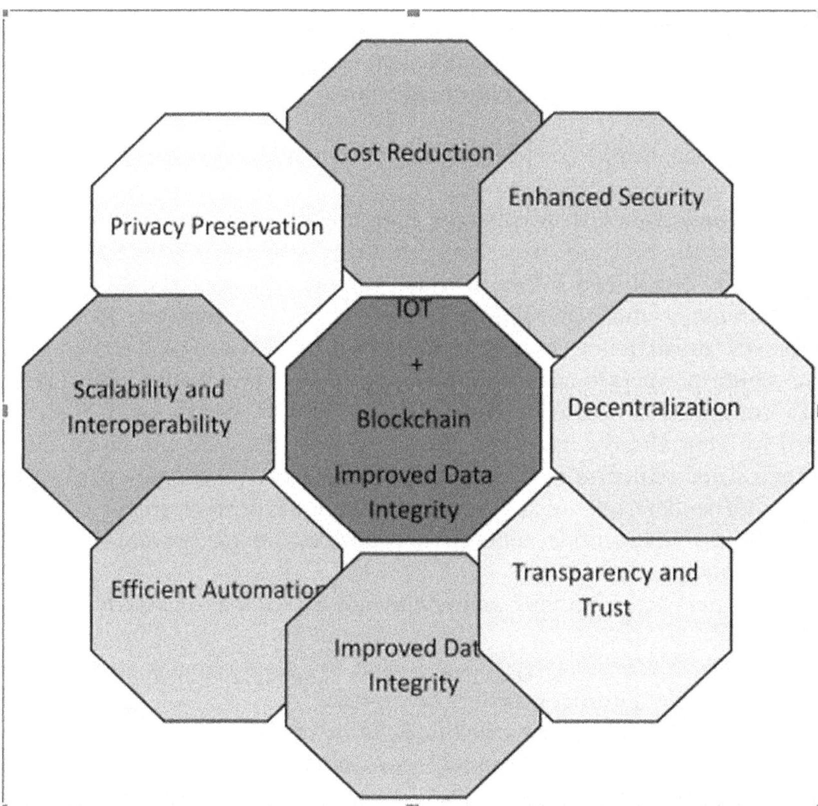

Figure 12.3 Benefits of Blockchain and IoT integration.

Decentralization: This lowers the chance of extensive cyberattacks by removing single points of failure.

Authentication: By limiting unwanted access, cryptographic keys guarantee safe device authentication. End-to-end encryption safeguards information while it's being transferred between IoT devices.*Auditability*: Safe tracking and verification are made possible by the Blockchain's visible recording of every action. All of these characteristics work together to reduce security threats in IoT networks.

Decentralization: By eliminating single points of failure, decentralization reduces the likelihood of large-scale cyberattacks. Decentralization in Blockchain refers to the sharing of data and control among a network of nodes rather than being governed by a single central authority. In Internet of Things systems, this ensures the following:

Elimination of single points of failure: No single server or entity controls the network, reducing risks from outages or cyberattacks.

Enhanced resilience: If one node fails, the system continues to operate unaffected.

Distributed trust: Trust is established across all participants using consensus mechanisms, removing reliance on intermediaries.

This decentralization ensures robust, secure, and trustworthy IoT systems.

Transparency and trust: By offering auditable data trails, immutable transaction records promote trust between users and devices. By keeping an immutable, distributed ledger in which each transaction or data exchange is documented and accessible to all authorized participants, Blockchain improves transparency. This guarantees that all parties involved can audit and confirm operations without the need for middlemen. Blockchain uses consensus processes to validate transactions, doing away with the need for centralized control and fostering trust. Because the Blockchain's decentralized structure guards against manipulation and guarantees equitable, independent operation, users can have faith in the system's integrity. Trustworthy interactions between IoT devices are further automated by smart contracts.

This fosters accountability and confidence across IoT ecosystems.

Improved data integrity: Ensures that IoT data remains accurate and unaltered during transfers or storage.

Improved data integrity in blockchain for IoT: Data integrity refers to ensuring the accuracy, consistency, and reliability of data throughout its lifecycle. Blockchain enhances data integrity in IoT systems through the following mechanisms:

Immutable ledger: Blockchain records data in a distributed and tamper-proof ledger. Once data is added, it cannot be altered without consensus from the network, ensuring its reliability.

Cryptographic hashing: Data is secured with cryptographic hashes, allowing any unauthorized modification to be immediately detected.

Traceability: The Blockchain maintains a transparent and permanent history of all transactions, enabling thorough audits and provenance tracking.

Decentralized validation: Data is verified by multiple network participants (nodes) before being added to the Blockchain, preventing erroneous or malicious entries.

In IoT systems, where large amounts of data are continuously generated and transmitted, Blockchain's ability to ensure data integrity is critical. This prevents false data injection, mitigates risks of fraud, and builds user trust in the system.

Efficient automation: Smart contracts facilitate automatic execution of rules and agreements, streamlining IoT operations like device-to-device communication or financial transactions.

Blockchain enables efficient automation in IoT systems primarily through **smart contracts**, which are self-executing agreements encoded on the Blockchain. Key benefits include:

Streamlined operations: Smart contracts automate tasks like device-to-device communication, data sharing, and payment processing, reducing the need for manual intervention.

Reduced delays: Automation removes intermediaries, speeding up processes such as access control, supply chain tracking, or energy trading.

Cost efficiency: Eliminates administrative overheads by executing predefined rules automatically and accurately.

This automation enhances efficiency, scalability, and reliability in IoT networks.

Scalability and interoperability: Blockchain enhances scalability by managing distributed IoT networks and improving interoperability through standardized frameworks.

Scalability: Blockchain facilitates the management of large-scale IoT networks by distributing data processing across nodes, enabling the addition of new devices without overwhelming the system.

Layer-2 solutions (e.g., off-chain computation) and sharding further improve scalability by reducing Blockchain congestion.

Interoperability: IoT networks often involve heterogeneous devices using different protocols. Blockchain bridges this gap by providing a unified framework for communication and data sharing across diverse devices. Standardized APIs and protocols enhance compatibility and enable seamless interactions between devices, platforms, and networks.

This ensures smooth, efficient, and scalable IoT operations.

Privacy preservation: Users retain control over their data, with private Blockchain options or selective disclosure through zero-knowledge proofs. Blockchain enhances privacy in IoT systems by providing mechanisms to secure sensitive data and give users control over their information:

Decentralized identity management: Users can manage their identities directly without involving third parties, protecting personal data.

Encryption: Data stored on the Blockchain is encrypted, ensuring only authorized parties can access it.

Selective data disclosure: Privacy-preserving techniques like zero-knowledge proofs allow users to verify claims without revealing the underlying data.

Secure access control: Smart contracts enforce strict access policies, limiting data exposure.

These features ensure robust privacy protection for IoT applications.

Cost reduction: Reduces intermediaries and associated costs by enabling direct interactions between IoT devices. Blockchain reduces costs in IoT systems by eliminating intermediaries, automating processes, and improving efficiency. Key factors include:

Elimination of intermediaries: Blockchain enables direct device-to-device transactions, removing the need for service providers.

Process automation: Smart contracts reduce operational overhead by automating repetitive tasks like data validation, billing, or compliance checks.

Improved resource management: Decentralized data storage and computation optimize resource utilization, minimizing infrastructure costs.

These features make IoT networks more cost-efficient while maintaining security and scalability.

12.5 APPLICATIONS OF BLOCKCHAIN-DRIVEN IOT

Applications for Blockchain and IoT integration are numerous and provide improved automation, security, and transparency in a variety of sectors. Here is a thorough explanation of the main usage cases.

Supply chain management: Blockchain guarantees supply chain traceability and end-to-end visibility. Blockchain securely stores the location, temperature, and other data that is gathered by IoT devices like RFID and sensors. Benefits include less fraud, guaranteed product authenticity, and effective cross-border tracking of items. For instance, keeping track of perishable items (like food or vaccinations) to keep an eye on conditions and guarantee quality compliance.

Smart cities: IoT devices in smart cities use Blockchain technology to manage infrastructure, garbage, traffic, and utilities. Devices can share data securely, which enhances urban operations. This makes decentralized energy networks, automatic parking systems, and real-time traffic monitoring possible. For instance, Blockchain-stored IoT data is used to dynamically alter smart traffic signals for security and transparency.

Energy trading: Peer-to-peer (P2P) energy trading that uses Blockchain and smart meters connected to the IoT. Without middlemen, homeowners can sell extra energy (from solar panels, for example) to other people. Blockchain guarantees energy transactions that are safe, transparent, and automated. An example would be a decentralized energy market where users may safely exchange energy credits.

Healthcare: Blockchain securely saves health data generated by wearable sensors, to guard against manipulation or unwanted access. This provides for the private exchange of patient data between insurance, physicians, and hospitals. For instance, Blockchain ensures tamper-proof logs for diagnostics and insurance claims by storing IoT-collected data from insulin pumps.

Industrial IoT (IIoT): IoT devices monitor operations, inventories, and machine performance in production. Secure and unchangeable records are guaranteed by Blockchain. Benefits include preventing data manipulation, automating part replacement based on IoT signals, and streamlining quality assurance. For instance, a Blockchain ledger that tracks IoT sensor data and machine maintenance logs to anticipate faults and save downtime.

Asset tracking: Blockchain and IoT monitor the whereabouts and state of assets such as machines, cars, and cargo. Benefits include increased security, less asset theft, and accountability. For instance, fleet management systems that use Blockchain to encrypt and distribute data to stakeholders and IoT devices for tracking vehicles.

Overview of agriculture: Blockchain collects data from IoT devices that monitor soil conditions, weather, and crop health in order to improve farming operations. Benefits: Ensures food traceability from farm to fork and helps optimize resource use. A Blockchain system that monitors organic produce to ensure standards are fulfilled would be one example.

Logistics and transportation overview: Real-time data (such as location and fuel consumption) is recorded by IoT-enabled cars and shipping containers and is kept on Blockchain for monitoring and analysis. Benefits: Boosts transparency, guarantees safe payment processing, and cuts down on shipment delays. For instance, Blockchain-based tracking helps expedite customs clearance for international shipping.

Autonomous vehicles: Blockchain guarantees safe connection between vehicles and infrastructure, while IoT sensors in autonomous cars produce a lot of operational data. Benefits include preventing hacking, boosting trust, and facilitating easy data exchange between automobiles. For instance, vehicle-to-vehicle communication to prevent collisions or automated toll payments.

Insurance description: To streamline claims processing, Blockchain securely saves data collected by IoT devices about insured assets (such as homes and vehicles). Benefits: Quickens claims settlement, guarantees openness, and lowers fraud. For instance, Blockchain-verified insurance claims are automatically triggered by IoT-enabled flood sensors.

These uses demonstrate how Blockchain and IoT complement each other to improve security, openness, and productivity in a variety of sectors.

12.6 CHALLENGES AND LIMITATIONS

There are many challenges to overcome in order to successfully integrate Blockchain technology with the IoT. Scalability, energy consumption, interoperability, cost, and regulatory frameworks are the main issues at hand.

12.6.1 Scalability

- *Transaction volume*: Because traditional Blockchains are unable to handle the volume of transactions generated by IoT devices, congestion and slower processing times arise.
- *Data storage*: The massive amount of data generated by IoT devices could quickly surpass Blockchain storage capacity, especially for devices with constrained resources.

12.6.2 Energy utilization

- *Mining*: For extensive IoT installations, the energy-intensive procedure of mining new blocks in conventional Blockchains may not be viable.
- *Consensus processes*: Some consensus processes use a lot of processing power, which might shorten battery life of IoT devices.

12.6.3 Interoperability

- *Communication protocols*: It is challenging to guarantee smooth connection between Blockchain networks and IoT devices due to their diverse use of communication protocols.
- *Data format*: Incompatible data formats used by various IoT devices and Blockchain platforms might make data analysis and sharing difficult.

12.6.4 Cost

- *Transaction fees*: Some Blockchains have expensive transaction fees that may prevent resource-constrained IoT devices from participating.
- *Costs of development*: Including Blockchain technology in IoT systems can be costly and complicated, requiring specific knowledge and infrastructure.

12.6.5 Frameworks for regulation

- *Lack of standards*: For Blockchain-based IoT applications, the lack of defined regulatory frameworks breeds ambiguity and prevents broad adoption.
- *Privacy issues*: Due to decentralized structure of Blockchain technology, sensitive IoT applications are particularly vulnerable to privacy and security issues.

Other difficulties include:

- *Security*: IoT devices and Blockchain networks must be secure since any weaknesses could be used by hostile parties.
- *Latency*: For real-time Internet of Things applications that need minimal latency, the intrinsic delay of Blockchain transactions can be a major problem.
- *Forking*: IoT applications that depend on the network's consistency and stability may be affected by the potential for Blockchain forks.

12.7 ADDRESSING CHALLENGES IN BLOCKCHAIN–IOT INTEGRATION

Despite its potential, the combination of IoT and Blockchain technology poses a number of serious obstacles. Nevertheless, creative fixes and technological developments are being investigated to get over these obstacles:

12.7.1 Scalability

- *Layer-2 Solutions*: The overloaded Blockchain can be greatly relieved by shifting computations and non-essential data to state channels or sidechains.
- *Sharding*: Transaction processing speed and scalability can be increased by splitting the Blockchain into smaller, easier-to-manage shards.
- *Distributed ledger technologies* (DLTs): Investigating substitute DLTs that provide greater scalability and quicker transaction confirmation times, such as Directed Acyclic Graphs (DAGs).

12.7.2 Energy consumption

- *Proof-of-Stake* (PoS): replacing Proof-of-Work (PoW) with PoS consensus processes, which use less energy and processing capacity.
- *Energy-efficient hardware*: Making use of specialized hardware, including ASICs with reduced power usage, made for Blockchain mining.
- *Off-chain processing*: To lessen the Blockchain's processing burden, non-essential data can be processed and stored off-chain.

12.7.3 Interoperability

- *Standardized protocols*: Creating standardized protocols to help Blockchain networks and IoT devices coordinate over data use.
- *Interoperability platforms*: Developing platforms that can connect various IoT and Blockchain standards.
- *Data bridges*: Putting data bridges into practice to convert data between various protocols and formats.

12.7.4 Cost

- *Layer-2 solutions*: To save transaction costs and boost productivity, use layer-2 solutions.
- *Public–private partnerships*: Working together with businesses and governments to build public infrastructure and cut expenses.

- *Token economy*: Creating token economies to reward network contributions and encourage involvement.

12.7.5 Security

- *Cryptographic techniques*: Using robust cryptographic techniques to protect data storage and transfer.
- *Secure hardware modules* (SHMs): Secure device identities and sensitive cryptographic keys are safeguarded by SHMs.
- *Regular security audits*: Evaluate possible security risks, conduct frequent vulnerability assessments and resolve the risks.

12.7.6 Privacy

- Zero-knowledge proofs are a technique for confirming information without disclosing private information.
- Homomorphic encryption is encryption that permits calculations on the encrypted data eliminating the requirement of decrypting it.
- *Privacy-preserving protocols*: Putting privacy-preserving protocols into place to safeguard sensitive data and user information.

12.7.7 The regulatory framework

- *Industry collaboration*: Working together with regulators and industry stakeholders to create regulations that are uniform and unambiguous.
- *Self-regulation*: Creating industry norms and best practices to guarantee moral and responsible Blockchain and Internet of Things usage.
- *Global collaboration*: Promoting global collaboration to standardize regulatory structures and ease cross-border data sharing.

Blockchain and IoT integration has the potential to unleash substantial value and bring about revolutionary change in a number of industries by tackling these issues and utilizing creative solutions.

12.8 CONCLUSION

To sum up, Blockchain in IoT shows how substantially Blockchain technology may impact IoT systems. Structured Blockchain adoption will be essential to enhancing the security, dependability, and effectiveness of these systems as the IoT continues to develop across all industries. This creates intriguing avenues for further study. Setting baseline security standards for IoT software that can manage the high computational demands of Blockchain integration, making sure smart contracts are fully validated before being used in Blockchain–IoT systems, and developing more flexible governance models are important topics. Investigating the legal and regulatory ramifications of Blockchain use is also crucial, as is creating precise standards and protocols that encourage compatibility across various IoT applications and devices. Also crucial are improving data validation methods for Blockchain-based IoT transactions and encouraging closer cooperation between IoT ecosystem participants. By addressing these issues, researchers and business experts will be able to create Blockchain–IoT systems that are safe, open, and incredibly effective.

REFERENCES

1. Ayoade G, Karande V, Khan L, Hamlen K. Decentralized IoT data management using blockchain and trusted execution environment. In *2018 IEEE International Conference on Information Reuse and Integration (IRI)* 2018 Jul 6 (pp. 15–22). IEEE.
2. Shalini P. Blockchain technology: Architecture, consensus, and future trends. *Emperor Journal of Applied Scientific Research*. 2024, 6(3):73–84.
3. Alam T. Blockchain-based Internet of Things: Review, current trends, applications, and future challenges. *Computers*. 2023, 12:6 [Internet]. 2022.
4. Makhdoom I, Zhou I, Abolhasan M, Lipman J, Ni W. PrivySharing: A blockchain-based framework for privacy-preserving and secure data sharing in smart cities. *Computers & Security*. 2020 Jan 1;88:101653.
5. Bhan R, Pamula R, Faruki P, Gajrani J. Blockchain-enabled secure and efficient data sharing scheme for trust management in healthcare smartphone network. *The Journal of Supercomputing*. 2023 Sep;79(14):16233–74.
6. Wang Z, Chen Q, Liu L. Permissioned blockchain-based secure and privacy-preserving data sharing protocol. *IEEE Internet of Things Journal*. 2023 Feb 7;10(12):10698–707.
7. Banerjee M, Lee J, Choo, KR. A blockchain future for Internet of Things security: A position paper. *Digital Communications and Networks. 2018;4(3):149–160.*
8. Uddin MA, Stranieri A, Gondal I, Balasubramanian V. A survey on the adoption of blockchain in IoT: Challenges and solutions. *Blockchain: Research and Applications*. 2021 Jun 1;2(2):100006.
9. Ali MS, Vecchio M, Pincheira M, Dolui K, Antonelli F, Rehmani MH. Applications of blockchains in the Internet of Things: A comprehensive survey. *IEEE Communications Surveys & Tutorials*. 2018 Dec 18;21(2):1676–717.
10. Issa W, Moustafa N, Turnbull B, Sohrabi N, Tari Z. Blockchain-based federated learning for securing Internet of Things: A comprehensive survey. *ACM Computing Surveys*. 2023 Jan 13;55(9):1–43.
11. Dwivedi SK, Roy P, Karda C, Agrawal S, Amin R. Blockchain-based Internet of Things and industrial IoT: A comprehensive survey. *Security and Communication Networks*. 2021;2021(1):7142048.

12. Zyskind G, Nathan O. Decentralizing privacy: Using blockchain to protect personal data. In *2015 IEEE Security and Privacy Workshops* 2015 May 21 (pp. 180–184). IEEE.
13. Vashisht S, Gaba S, Dahiya S, Kaushik K. Security and privacy issues in IoT systems using blockchain. In *Sustainable and Advanced Applications of Blockchain in Smart Computational Technologies* 2022 Sep 30 (pp. 113–127). Chapman and Hall/CRC.
14. Raza U, Kulkarni P, Sooriyabandara M. Low power wide area networks: An overview. *IEEE Communications Surveys & Tutorials*. 2017;19(2):855–73.
15. Atzori L, Iera A, Morabito G. The Internet of Things: A survey. *Computer Networks*. 2010;54(15):2787–805.
16. Tran NK, Babar A, Boan, J. Integrating blockchain and Internet of Things systems: A systematic review on objectives and designs. Journal of Network and Computer Applications. 2021;173:102844–102868.

Index

T

For Product Safety Concerns and Information please contact our EU
representative GPSR@taylorandfrancis.com
Taylor & Francis Verlag GmbH, Kaufingerstraße 24, 80331 München, Germany

www.ingramcontent.com/pod-product-compliance
Ingram Content Group UK Ltd.
Pitfield, Milton Keynes, MK11 3LW, UK
UKHW022218190926
475702UK00003B/428
* 9 7 8 1 0 3 2 9 5 4 0 8 0 *